T0295101

HEALTHCARE TECHNOLOGIES SERIES 48

Blockchain Technology in e-Healthcare Management

IET Book Series on e-Health Technologies

Book Series Editor: Professor Joel J.P.C. Rodrigues, College of Computer Science and Technology, China University of Petroleum (East China), Qingdao, China; Senac Faculty of Ceará, Fortaleza-CE, Brazil and Instituto de Telecomunicações, Portugal

Book Series Advisor: Professor Pranjal Chandra, School of Biochemical Engineering, Indian Institute of Technology (BHU), Varanasi, India

While the demographic shifts in populations display significant socio-economic challenges, they trigger opportunities for innovations in e-Health, m-Health, precision and personalized medicine, robotics, sensing, the Internet of things, cloud computing, big data, software defined networks, and network function virtualization. Their integration is however associated with many technological, ethical, legal, social, and security issues. This book series aims to disseminate recent advances for e-health technologies to improve healthcare and people's wellbeing.

Could you be our next author?

Topics considered include intelligent e-Health systems, electronic health records, ICT-enabled personal health systems, mobile and cloud computing for e-Health, health monitoring, precision and personalized health, robotics for e-Health, security and privacy in e-Health, ambient assisted living, telemedicine, big data and IoT for e-Health, and more.

Proposals for coherently integrated international multi-authored edited or co-authored handbooks and research monographs will be considered for this book series. Each proposal will be reviewed by the book Series Editor with additional external reviews from independent reviewers.

To download our proposal form or find out more information about publishing with us, please visit https://www.theiet.org/publishing/publishing-with-iet-books/.

Please e-mail your completed book proposal for the IET Book Series on e-Health Technologies to: Amber Thomas at athomas@theiet.org or author_support@theiet.org.

Blockchain Technology in e-Healthcare Management

Edited by
Suyel Namasudra and Victor Hugo C. de Albuquerque

The Institution of Engineering and Technology

Published by The Institution of Engineering and Technology, London, United Kingdom

The Institution of Engineering and Technology is registered as a Charity in England & Wales (no. 211014) and Scotland (no. SC038698).

First published 2022

The Institution of Engineering and Technology
Futures Place
Kings Way, Stevenage
Hertfordshire SG1 2UA, United Kingdom

www.theiet.org

British Library Cataloguing in Publication Data
A catalogue record for this product is available from the British Library

ISBN 978-1-83953-602-1 (hardback)
ISBN 978-1-83953-603-8 (PDF)

Typeset in India by MPS Limited

Cover Image: Luis Alvarez / DigitalVision via Getty Images

Contents

About the Editors

Suyel Namasudra has received Ph.D. degree from the National Institute of Technology Silchar, Assam, India. He was a post-doctorate fellow at the International University of La Rioja (UNIR), Spain. Currently, Dr. Namasudra is working as an assistant professor in the Department of Computer Science and Engineering at the National Institute of Technology Agartala, Tripura, India. Before joining the National Institute of Technology Agartala, Dr. Namasudra was an assistant professor in the Department of Computer Science and Engineering at the National Institute of Technology Patna, Bihar, India. His research interests include blockchain technology, cloud computing, IoT, and DNA computing. Dr. Namasudra has edited 4 books, 5 patents, and 70 publications in conference proceedings, book chapters, and refereed journals like IEEE TII, IEEE T-ITS, IEEE TSC, IEEE TCSS, IEEE TCBB, ACM TOMM, ACM TOSN, ACM TALLIP, FGCS, CAEE, and many more. He has served as a Lead Guest Editor/Guest Editor in many reputed journals like ACM TOMM (ACM, IF: 3.144), MONE (Springer, IF: 3.426), CAEE (Elsevier, IF: 3.818), CAIS (Springer, IF: 4.927), CMC (Tech Science Press, IF: 3.772), Sensors (MDPI, IF: 3.576), and many more. Dr. Namasudra has participated in many international conferences as an organizer and session Chair. He is a member of IEEE, ACM, and IEI. Dr. Namasudra has been featured in the list of the top 2% scientists in the world in 2021 and 2022, and his h-index is 27.

Victor Hugo C. de Albuquerque (Senior Member of IEEE) is currently a Professor and Senior Researcher at the Department of Teleinformatics Engineering (DETI)/Graduate Program in Teleinformatics Engineering (PPGETI) at the Federal University of Ceará (UFC), Brazil. He earned a Ph.D in Mechanical Engineering from the Federal University of Paraíba (UFPB, 2010), a MSc in Teleinformatics Engineering from the PPGETI/UFC (UFC, 2007). He completed a BSE in Mechatronics Engineering at the Federal Center of Technological Education of Ceará (CEFETCE, 2006). He has experience in Biomedical Science and Engineering, mainly in the research fields of: Applied Computing, Intelligent Systems, Visualization and Interaction, with specific interest in Pattern Recognition, Artificial Intelligence, Image Processing and Analysis, as well as Automation with respect to biological signal/image processing, biomedical circuits and human/brain-machine interaction, including Augmented and

Virtual Reality Simulation Modeling for animals and humans. Prof. Victor is a full Member of the Brazilian Society of Biomedical Engineering (SBEB). He is Editor-in-Chief of the Journal of Biomedical and Biological Sciences, and, also, of the Journal of Artificial Intelligence and Systems, and Journal of Biological Sciences, as well as Associate Editor of the IEEE Journal of Biomedical and Health Informatics; Computers in Biology and Medicine; Frontiers in Cardiovascular Medicine; Computational Physiology and Medicine; Applied Soft Computing; IEEE Access, Frontiers in Communications and Networks, Computational Intelligence and Neuroscience, Measurement, IET Quantum Communication, and he has been Lead Guest Editor of several high-reputed journals, and TPC member of many international conferences.

Preface

Nowadays, approx. 2.5 quintillion bytes of data are produced every day and most of the data are communicated over the internet. As there are many hackers and malicious users over the Internet, security and trust issues are very critical. Blockchain is a novel technique to solve these issues, which allows a radical way of executing transactions among several entities, such as businesses, individuals, and machines. Blockchain can be defined as a Distributed Ledger Technology (DLT) that secures and records transactions in a Peer to Peer (P2P) network instead of using single or many servers. Here, each record is saved on many interconnected systems, which keep identical information. In a blockchain network, numerous transactions are grouped into several blocks, and each block is linked to the previous block. Information of a blockchain network is immutably recorded across a P2P network by each block. Blockchain has many applications, such as healthcare, finance, Internet of Things (IoT), data storage, decentralized cryptocurrency, and many more.

In the healthcare sector, e-healthcare is very important in which Information and Communication Technology (ICT) tools are used to maintain the health information of any patient. Currently, healthcare records are saved in databases controlled by an individual user, organization, or large groups of organizations. As there are numerous malicious users, this information must not be shared with other organizations due to security issues and the chances of the data being modified or tampered. Blockchain technology can be used in e-healthcare systems to securely exchange healthcare data, which can be accessed by organizations sharing the same network, allowing doctors/practitioners to provide better care for patients. The key properties of blockchain technology, such as immutability and transparency, improve e-healthcare interoperability.

This edited book discusses many applications of blockchain technology in e-healthcare management, including applications in the pharmaceutical industry, healthcare data security, and many more. Chapter 1 of this edited book discusses the fundamentals of blockchain technology. Chapter 2 represents how blockchain technology can be used to manage healthcare records, while Chapter 3 deliberates upon applications of blockchain technology in the pharmaceutical sector. Chapter 4 focuses on many techniques that are being used to manage health insurance using blockchain technology. Nowadays, blockchain technology is continuously used in various ways to improve e-healthcare data security, which are discussed in Chapter 5. Chapter 6 deals with the integration of blockchain technology with IoT to monitor a patient's health. After that, Chapter 7 represents a case study of blockchain technology to manage e-healthcare data. Finally, challenges and future work directions in e-healthcare using blockchain technology are represented in Chapter 8.

Suyel Namasudra and Victor Hugo C. de Albuquerque
Editors

Chapter 1

Blockchain technology: fundamentals, applications, and challenges

Sangjukta Das[1], Suyel Namasudra[2] and Victor Hugo C. de Albuquerque[3]

Abstract

Blockchain, also known as distributed ledger technology, has gained widespread adoption in many fields apart from financial transactions. With high security and transparency, it is used in supply chain management systems, healthcare, payments, business, Internet of Things (IoT), voting systems, and many more for securing a wide range of digital assets from tampering. Blockchain technology is also used to share, replicate, and synchronize data, and to track transactions and assets in any type of business across different geographical locations. Rather than working with a central authority to control all the data transactions, blockchain technology uses a consensus mechanism, where all the nodes of the system need to agree on a new transaction. At first, this chapter discusses all the fundamental aspects and technical details along with the advantages and disadvantages of blockchain. It also discusses different types of blockchain networks with their pros and cons, smart contracts, and hyperledger technology. Then, several applications of blockchain technology and challenges in adopting it in various industries are also given in this chapter to show how the technology works as a whole.

Keywords: Decentralized network; Merkle root; Consensus algorithm; Smart contract

1.1 Introduction

To bring revolution, any technology must have an ample amount of advantage over conventional work systems. Blockchain is one such technology, being adopted by

[1]Department of Computer Science and Engineering, National Institute of Technology Patna, Bihar, India
[2]Department of Computer Science and Engineering, National Institute of Technology Agartala, Tripura, India
[3]Federal University of Ceará, Brazil

almost every industry for record-keeping. In traditional methods, participants on a network keep their ledgers and other records for recording transactions and tracking assets. Sometimes, this method can be expensive and time-consuming because it involves mediators that may charge a service fee, and also, multiple copies of records for multiple participants may result in a processing delay. Due to this, working in traditional systems is painful, while executing agreements and maintaining numerous records in multiple ledgers. For example, while purchasing a house, one probably needs to sign a substantial amount of paperwork from numerous parties for the transaction to be completed. Similarly, registering a vehicle or tracking medical records can be awful and challenging for an individual. Moreover, in a traditional system, if the central authority is compromised due to a cyberattack or fraud, the entire business network gets affected. Blockchain has the capability to fundamentally change how traditional systems work by redefining these procedures, as well as many others. This technology brings a major revolution by removing the need for a centralized trusted authority in a widely distributed network. Here, without relying on any centralized third party, trustworthy and secure transactions can be made across an untrusted network. In this type of network, all peers can interact with one another and have equal rights. Figure 1.1 depicts the pictorial difference between a centralized server system and a peer-to-peer network. In a blockchain network, based on a consensus algorithm, a transaction can be trusted as valid because it offers an immutable and enduring record of transactions, which cannot be modified or altered. Thus, blockchain technology is posing a threat to traditional systems and can be considered the technology of the future.

Figure 1.2 represents business networks before and after introducing blockchain technology. Figure 1.2(a) shows the business network before the inception of blockchain technology in which all the nodes have their own copy of the ledger, however, the central system has the "golden record," i.e., centralized ledger. Figure 1.2(b) shows the business network after adopting blockchain technology.

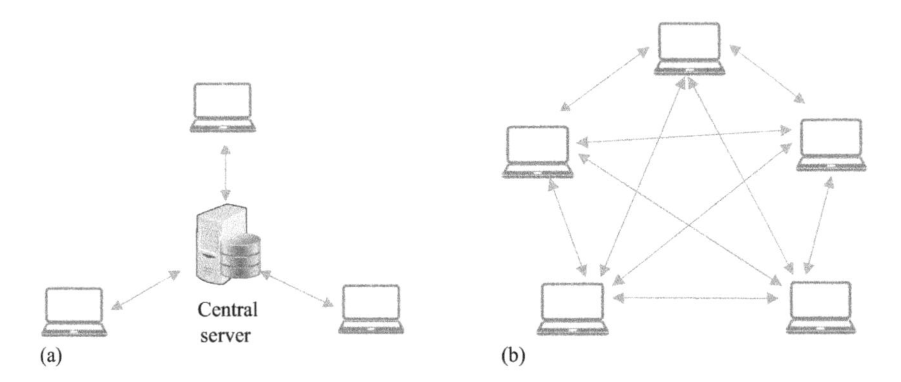

Figure 1.1 A simple block diagram of networks: (a) client-to-server network and (b) peer-to-peer network

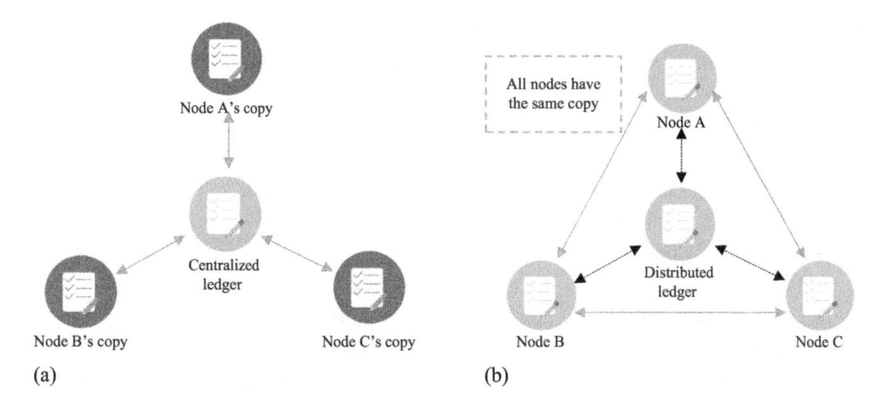

Figure 1.2 Business models: (a) before introducing blockchain and (b) after introducing blockchain

There is a single distributed ledger and all the nodes have the same level of access to it. The nodes or participants can share a ledger, which is updated through peer-to-peer replication whenever a transaction occurs. Each participant can send and receive transactions from other participants. It is worth noting that the data is replicated through a process called "consensus" process. Consensus means that a transaction is validated, when all the participants agree on its validity. Blockchain technology is less vulnerable because it uses consensus mechanisms to validate a transaction. It is also efficient and economical because it eliminates the effort of duplication and the need for mediators [1].

Blockchain provides the facility to record transactions and track assets in any business network. An asset can be anything, such as land, cash, house, copyrights, or patents. Anything that can add real value in exchange for something can be tracked and traded on a blockchain network. As the name suggests, blockchain is a chain of blocks, where transactional data are stored in blocks and these blocks are linked together to form a chain. Therefore, it is a chronological sequence of blocks that can record and confirm the time and sequence of a valid transaction record. Each new block is linked to the last block of the blockchain within a network governed by the rules agreed on by all network participants. The first block of the blockchain is called *Genesis Block* and the block preceded by a given block is called the parent block. The genesis block is the only block, which does not have a preceding block. Each block has two unique components called block header and block body [2]. The block header field of a block contains several components, such as block version, previous block hash, timestamp, nonce, Merkle root, and difficulty bit. Among these components, the most important is the Merkle root, which is a hash used to uniquely identify a block. The body of a block contains a transaction list that contains new transactions. Along with the actual transactions, the transaction list also contains several crucial components that are important for accepting a new transaction [3].

In the blockchain, any block B_2 is immutably linked to a single preceding block B_1 through a hash function $H(B_1)$. Any change in block B_1 would result in an invalid hash in B_2 and all following blocks. Therefore, to ensure the integrity of a block and the data contained in it, the block is digitally signed. The main features of blockchain technology are transparency, immutability, decentralization, disintermediation, redundancy, scalability, etc. Currently, it has many applications in the field of finance [4–7], transportation systems [8], Internet of Things (IoTs) [9,10], data storage [11,12], data provenance [13,14], decentralized cryptocurrency, identity [15–17], blockchain business, smart property, and many more.

The primary goal of this chapter is to introduce all fundamental elements of blockchain technology. This chapter has a detailed discussion on the structure of blocks, block generation, and the complete transaction execution mechanism. This chapter discusses numerous uses of blockchain technology in various domains in detail. Additionally, the challenges, advantages, and disadvantages of applying blockchain technology in many fields are also discussed, which can be quite helpful for researchers, academics, and business experts.

The remainder of the chapter is divided into the following sections. Section 1.2 discusses fundamentals of the blockchain technology including its history, characteristics, advantages, disadvantages, and different types. Section 1.3 deals with the technical details of the blockchain network. Some application use cases and challenges in the adoption of blockchain technology are discussed in Sections 1.4 and 1.5, respectively. Then, the conclusion of the chapter is given in Section 1.6.

1.2 Fundamentals of blockchain technology

In this section, the fundamentals of blockchain including its history, characteristics, advantages, disadvantages, and types of blockchain are discussed.

1.2.1 History

The popularity of blockchain technology has already made it a common name in the worldwide record-keeping system.

In 1982, cryptographer D.L. Chaum first proposed a blockchain-like protocol in a dissertation "Computer Systems established, maintained, and trusted" [18,19]. In 1991, for the first time, Stuart Haber and Scott Stornetta described a cryptographically secured chain of blocks [20,21]. In 1992, the Merkle tree's concept was incorporated into the design to improve efficiency by allowing many documents to be collected into one block [22]. In 1998, Nick Szabo designed a decentralized digital currency called "Bit gold" [23]. In 2004, cryptographer Hal Finney invented Reusable Proof-of-Work (RPoW) as a prototype for digital cash. It was a significant phase in the development of cryptocurrencies.

Next, in late 2008, blockchain was introduced by an unknown individual, under the pseudonym of "Satoshi Nakamoto," as a part of the proposal for the online cryptocurrency "Bitcoin." In November 2008, a paper titled "Bitcoin: A Peer-to-Peer Electronic Cash System," was posted on the cryptography and cryptography policy mailing list [24]. This paper listed the details of the Bitcoin protocol along with the original code for early versions of Bitcoin. Although the proposal received a response concerning scalability and realistic prospects for Bitcoin, initially. Later on, the success of Bitcoin made blockchain technology popular to facilitate trustworthy and secure transactions. It should be noted that Bitcoin and blockchain are not the same things. Blockchain is the digital foundation that supports applications, such as Bitcoin, which is the first use case of blockchain. However, blockchain has its extent far beyond Bitcoin.

The evaluation of blockchain technology from version 1.0 to version 3.0 can be considered as the amalgamation of many fields, such as network and security, finance and economy, cryptography, distributed networking, software engineering, and data structure. Satoshi Namakato published a cryptographically secured method using blockchain technology to solve the double spend situation. In double spending, an attacker spends the same digital currency more than once to get multiple numbers of services. Namakato defined blockchain technology as a data structure of the hashed chain of timestamps. Every timestamp contains the previous timestamp in its hash, making a chain of hash. All three versions of blockchain are described below:

1. *Version 1.0 (currency)*: Blockchain version 1.0 was first introduced by security researcher Hall Finley in 2005 with an aim to create money by solving computational puzzles. After that, many other cryptocurrencies based on blockchain technology were introduced to support financial transactions.
2. *Version 2.0 (smart contracts)*: Blockchain version 2.0 extended the technology beyond cryptocurrencies and introduced smart contracts, which are automated computer programs that execute whenever a required condition is satisfied. In this version, the blockchain was able to process a higher number of transactions on the public network.
3. *Version 3.0 (DApps)*: Blockchain version 3.0 introduces Decentralized Applications (DApps), which use decentralized storage and communication system. There are many decentralized applications or DApps, such as BitMessage, BitTorrent, Tor, and Popcorn.

The current version, i.e., version 4.0 (blockchain for the industry) introduces blockchain-enabled strategies to solve business or industrial requirements.

1.2.2 Characteristics

The key characteristics of blockchain are discussed in this subsection.

1. *Consensus:* Every blockchain network has a consensus algorithm, which is a decision-making algorithm. Here, to make a transaction valid, every node of that network must agree on its validity.
2. *Provenance:* Participants know where the asset came from and how its ownership has changed over time.
3. *Autonomy:* Every node of a blockchain network can access, transfer, and update data independently and autonomously without third-party intervention.
4. *Immutability:* In a blockchain network, the record of a transaction cannot be changed once it is recorded in the ledger.

5. *Decentralized:* In a blockchain, no central authority is responsible for looking after the network. Rather, a group of nodes maintains the network to make it decentralized.
6. *Faster settlement:* As compared to traditional systems, blockchain-enabled systems can provide a faster settlement.

1.2.3 Advantages

There are many advantages of blockchain technology. Some of them are mentioned below:

1. *Enhanced security:* Blockchain can significantly change the way critical information is stored. It helps to prevent fraud and unauthorized activity by creating a record that cannot be altered using security attacks.
2. *Greater transparency:* The use of distributed ledger makes transactions and data be recorded identically in multiple locations. Every participant of the network sees the same information at the same time, which provides full transparency.
3. *Instant traceability*: Blockchain technology is created in a way that it can leave an unalterable audit trail that makes it simple to access and trace any information from that chain.
4. *Increased efficiency and speed:* The traditional paper-based record-keeping systems are error-prone, cumbersome, and often require third-party intervention. However, blockchain-based systems can automate these processes by using smart contracts. Thus, transaction settlement can be done in a faster way and can be stored on the distributed ledger along with transaction details.
5. *Automation:* Transactions can be automated with smart contracts to increase processing efficiency and speed. In a blockchain network, these are programs, codes, or logic that can automate many tasks of the network.

1.2.4 Disadvantages

This subsection represents the disadvantages of blockchain technology.

1. *Scalability:* In a blockchain network, scalability indicates the time required to process and validate transactions. With the increasing number of nodes, the network bandwidth, overall storage space, and power consumption of the network become limited.
2. *Energy consumption:* Blockchain uses encryption technology to provide security and establish consensus algorithms over the distributed network. To add a transaction to the blockchain's distributed ledger, many complex processes are executed, which requires a large amount of power consumption.
3. *Slow and cumbersome:* As compared to traditional payment methods like debit cards or cash, blockchain technology is slow to execute any transaction due to its distributed nature, complexity, and encryption operation. Transactions of Bitcoin take an hour to finalize.

4. *Standardization:* There is no standardization in blockchain technology to evaluate its compatibility with current laws and regulations, as well as to integrate it with the existing systems. Lack of standardization and clarity restrict the adoption of this technology on a large scale.
5. *Inefficiency:* Multiple network users validating the same operation is inefficient and wasteful. Specifically, the use of the Proof-of-Work (PoW) consensus algorithm is completely inefficient because there is only one winner and every other miner's effort is wasted.
6. *Security:* Although blockchain technology provides high security, this does not mean that it is completely secure. The blockchain network can be corrupted in a number of ways, some of which are discussed below:
   - *51% Attack:* If a node has control over 51% of the total network nodes, a 51% attack can be occurred in that network. In this case, the ledger can be altered and double spending is also possible. On a network with controllable nodes or miners, this attack is more likely to occur [25].
   - *Double-spending:* Double-spending is only possible, if a network is vulnerable to a 51% attack. Here, the same digital token can be spent simultaneously more than once. However, the blockchain network employs Proof-of-Stake (PoS), PoW, and other techniques to stop double spending.
   - *DDoS attack:* In this attack, the nodes are flooded with a large number of similar requests.
   - *Breaking blockchain's cryptography:* The use of quantum computing can crack blockchain's cryptographic algorithm. An attacker can break the private key assigned to each node, if the randomness of the key is low.

1.2.5 Types

There are four different types of blockchain networks depending on demand, preference, and priority. A brief description of the four different types of blockchain networks is given below:

1. *Public blockchain:* A public blockchain is permissionless and entirely decentralized in nature. The major purpose of a public blockchain network is to create an open environment, where anyone can join and contribute to the network. Here, any participant or node can participate in the core activities of the blockchain network and has equal rights to access the network and the ability to create and validate new blocks of data. Public blockchains are mostly used for exchanging and mining cryptocurrencies. A few benefits of using public blockchain are:
   - *No trust issues:* As everything is recorded on a distributed public ledger, which is public and immutable, there is no need to set up trusted authority separately.
   - *Security:* The decentralized nature makes the network more secure. The more nodes join the network, the more difficult it is for an attacker to hack the network.

- *Anonymity:* In a public blockchain network, every node's identity is anonymous, which ensures that no node can be traced by analyzing some transactions.
- *Immutability:* The pubic blockchain is immutable. Thus, no one can modify the data once it is recorded in the network.
- *Transparency:* Transparency of a public blockchain network makes all transactions available to all nodes.

   All these benefits of a public blockchain network make it a valuable solution for completely decentralized, democratized, and authority-free functioning. However, there still exist several disadvantages in a public blockchain, which are discussed below:

- *Low speed:* The first major disadvantage of a public blockchain is extremely slow processing speed. For example, Bitcoin processes only seven transactions per second.
- *Lack of scalability:* Another disadvantage of a public blockchain network is scalability. As the network grows larger, the network bandwidth, storage space, and power supply become limited.
- *Energy consumption:* Public blockchain requires a large amount of energy to operate.
- *Chances of fraudulent activities:* Since anyone can join a public blockchain, it may include fraudulent participants engaged in malicious activity.

2. *Private blockchain:* It is a permissioned blockchain controlled by a central organization, which decides who can join the network. A node of the private blockchain cannot perform all the network activity and can see only those transactions it is authorized to view. This type of blockchain network can ensure a high degree of privacy as it requires authentication to access the data. Here, the consistency of the network data can be controlled more effectively. The advantages of using private blockchain are:
   - *High speed:* Unlike a public blockchain, private blockchain adds a limited number of nodes. Thus, it can process more transactions per second.
   - *More scalable:* As private blockchain has less number of nodes in the network, it can provide more scalability to the network.
   - *Highly efficient:* The network becomes more efficient due to limited resource requirements.
   - *Fully safe:* Due to the restricted access and permissioned relationships, a private blockchain is completely safe.
3. *Consortium blockchain*: It is a hybrid form of private and public blockchain networks, which overcomes the disadvantages of earlier types of blockchain networks. A consortium blockchain is a permissioned one, where a group of privileged nodes can take decisions for the overall benefit of the network. This overcomes a drawback of a private network by removing a single central authority and it is more decentralized than the private network. One example of a consortium blockchain network is hyperledger. The advantages of using consortium blockchain networks are faster processing speed, scalability, low transaction cost, low energy consumption, and prevention from malicious activity. A comparison among public, private, and consortium blockchain networks is shown in Table 1.1.

Table 1.1 Comparison of different types of blockchain

Properties	Public	Private	Consortium
Efficiency	Low	High	High
Immutability	Almost impossible to break	Could be broken	Could be broken
Centralized	No	Yes	Partial
Ownership	No one	Single node	Multiple nodes
Consensus process	Permissionless	Permissioned	Permissioned
Cost	High	Medium	Low
Speed	Slow	Fast	Fast
Network type	Decentralized	Partially decentralized	Partially decentralized or centralized

1.3 Technical details of blockchain technology

Blockchain uses distributed ledger technology that provides a unique way to securely record and transfer information even through an unsecured channel. All the information of a blockchain network is stored across multiple computers also called "nodes" of the network. Information added to the ledger is organized into blocks on the blockchain. As these blocks can store only a limited amount of information, new blocks are continually added to the ledger, forming a chain. In this section, all the technical details behind the working of blockchain technology are discussed.

1.3.1 Block

The term "Blockchain" implies that it is a chain of blocks. It is worth describing the components of this block and how the chaining of blocks is formed.

A block is a data structure comprising a few fields as shown in Figure 1.3. Apart from the block header and transaction list, there are other fields like block size and transaction counter. In a block, the block header contains information about the block itself. The transaction list of a block contains new transactions. The block size is consistent for the entire network. A transaction counter is a counter for the number of transactions in each block. The structure of a block is shown in Table 1.2, where the block header takes up to 80 bytes and the average size of a transaction is at least 250 bytes.

The block header contains:

- *Previous block hash*: It is the 256-bit hash value of the previous block.
- *Nonce*: A nonce is a 4-byte random field that miners adjust for every hash calculation to solve a PoW mining puzzle.
- *Timestamp*: It is the creation time of the current block.
- *Block version*: Block version is the version number of the block, which follows the block validation rules.

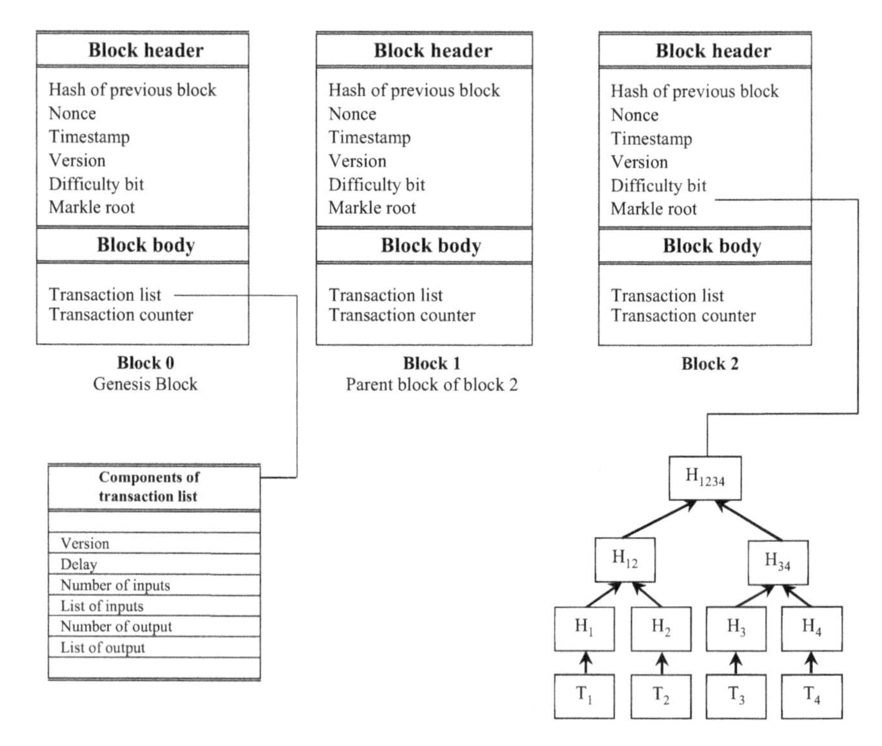

Figure 1.3 Structure of a block in a blockchain

Table 1.2 Structure of a block

Field	Description	Size in bytes
Block size	The size of the block	4
Block header	Block header contains block metadata	80
Transaction counter	Number of transactions in a block	1–9
Transaction list	Transactions recorded in a block	Variable size

- *Difficulty bit*: It is used to determine the difficulty of the PoW puzzle. The lower the value of it, the harder it is to generate a block.
- *Merkle root*: It is the hash value of the Merkle tree root that uniquely identifies a block.

A user shares transactions with some of the nodes of the blockchain network to submit transactions to the ledger. Then, these transactions are further shared with the other nodes of the system. Next, a specific node called the mining node appends these transactions to the blockchain. Mining nodes are a group of nodes responsible for broadcasting new blocks, appending transactions to the block, verifying those blocks, and finally, adding blocks to the blockchain. Thus, a block is a collection of authenticated transactions.

In simple words, a transaction is a record that records the transfer of any asset from one party to another. In a block, a transaction list contains several transactions of data. The components of transaction lists are version, lock time delay, and list of input and output.

- Version indicates the version number of the protocol being used. It is used to define the transactions.
- Lock time delay indicates the time after which a transaction can be accepted for a block.
- The list of inputs contains all the transactions accepted in the block.
- The list of outputs ensures the transfer of bitcoins (in satoshi) from the sender to the receiver. Note that satoshi is the smallest unit of Bitcoin.

A complete transaction execution process on a blockchain network is shown in Figure 1.4. In a transaction, units consumed by the transaction are called the Inputs (In), and the units created by a transaction are called the Outputs (Out). In Figure 1.3, transaction list B contains one input and two outputs. It receives 6.12 BTC, which is further split into two parts: 5.10 BTC and 1.02 BTC. These are recorded on the blockchain as they are parts of a transaction.

1.3.2 Chaining of blocks

In a blockchain network, blocks are linked together cryptographically to create the blockchain. This chaining is done among the blocks, where each block includes the hash of the previous header block as shown in Figure 1.5.

In Figure 1.5, there are three blocks: Blocks 0, 1, and 2. Consider that all three blocks store some transaction data up to 1 MB. Bitcoin blocks can store up to 1 MB of data approximately. Block 0 stores three transactions that makeup to 1 MB of

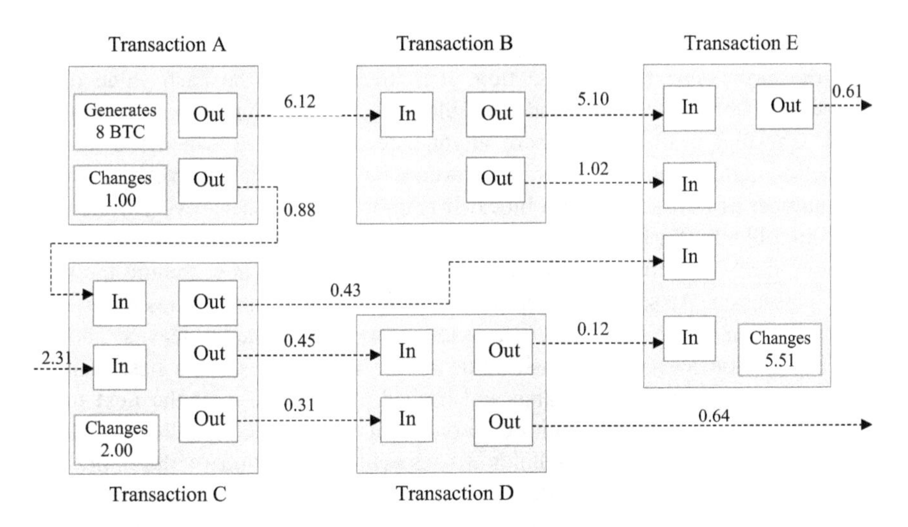

Figure 1.4 Transactions on the blockchain network

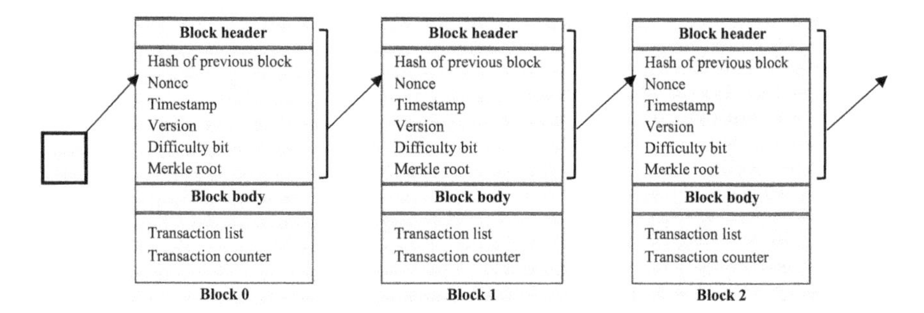

Figure 1.5 Chaining of blocks on the blockchain

data. The next transactions are stored in Block 1, making up to another 1 MB and this process goes on. These blocks now get a unique signature using hashing for the specific string of data in those blocks. Now, the blocks are linked together by adding the signature of Block 0 to Block 1. Here, the hash of the next block partially depends on the hash of the previous block because the next block contains the hash of the previous block. Any change inside a block changes the hash (signature). For example, any change in the data of Block 1 causes Block 2 to have a different hash. As a result, the hash added to Block 2 previously does not match the new hash anymore. Therefore, Blocks 1 and 2 are no longer chained to each other. This means changing a single block necessitates creating new signatures for all subsequent blocks until the end of the chain, which is practically impossible.

Note that the genesis block does not have a previous block, which means the previous block hash is set to zero.

1.3.3 Merkle tree

Merkle tree, also known as a binary hash tree, is a tree data structure that is used to store the hash value for a transaction. It is created from the hash value of the transactions stored in a block and it enables faster access during verification purposes. Consider a situation, where anyone needs to verify whether a specific transaction is stored on a block or not. It would be very inefficient to check through an N number of transactions in a block. It is impossible to check every transaction stored in a block of a blockchain network that contains millions of blocks. In this situation, the Merkle tree can provide speed and efficiency in searching the particular transaction. As shown in Figure 1.6, there are eight transactions recorded in the Merkle tree. In the Merkle tree, transactions are represented as leaves. The next higher-level nodes show the hash value of the transactions. This hash value is combined with another hash value and hashed again to abstract the next higher level. This process is repeated until only one hash is left, which is called the Merkle root. Finally, the highest level holds a hash with information of the entire tree. Notice that each level in a Merkle tree contains information about the level below.

Let us see how a Merkle root helps in searching a transaction. Consider that a node wants to find transaction 7 from the Merkle tree shown in Figure 1.6. Here,

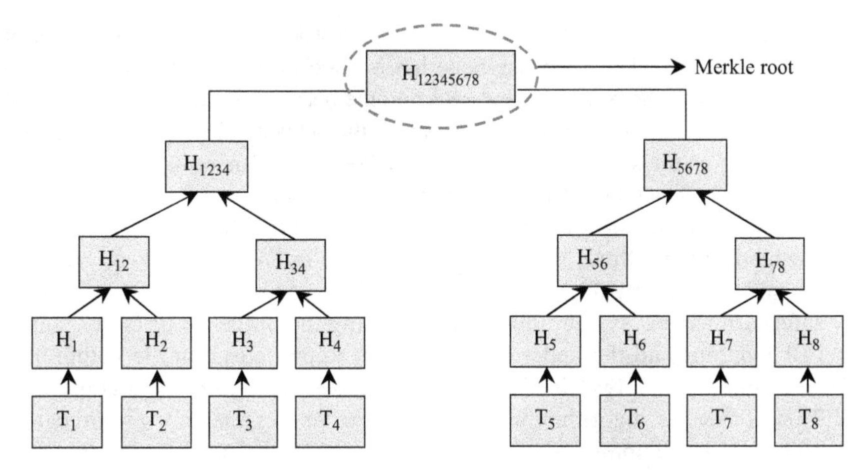

Figure 1.6 Constructing a Merkle tree

the Merkle root allows the node to skip half of the tree, which limits the search space between transaction 5 to 8. The node can reach transaction 7 in just three steps ($H_{12345678}$–H_{5678}–H_{78}–H_7) with the help of hashes. This takes comparatively fewer steps and time than searching through the whole tree. Here, the path to reach transaction 7 from the Merkle root is called a Merkle branch, which connects the root to a leaf.

1.3.4 Consensus algorithm

The blockchain validation process is also known as mining by some algorithms like PoW, PoS, etc. A consensus algorithm typically handles this validation process and provides the rules that the nodes must adhere to validate the blocks. If a block is verified by the validator node, then the block is added to the blockchain ledger as an immutable record. This implies that only the validator nodes can decide on the addition of a block to the blockchain [26]. Some of the main consensus algorithms are discussed below:

1. **Proof-of-Work:** In the proof-of-work consensus algorithm, the nodes or miners compete with each other to solve a cryptographic puzzle. The node that solves the puzzle first, gets the right to validate the block. To validate a block, the miners find a nonce to solve the cryptographic puzzle. Here, miners use brute force search to find the nonce that solves the puzzle, which makes the process time-consuming and mathematically challenging. Usually, the node or miner with more computational capacity solves the puzzle first. Some implementations of the PoW algorithm like Bitcoin offer a reward to the winner.
2. **Proof-of-Stake:** Proof-of-stake was initially designed to overcome the shortcomings of PoW. Here, block validators are selected based on the number of cryptocurrencies they have in the system. Block validators are selected often at random depending on how much cryptocurrency they are ready to stake to hold

the validation privileges. For example, if a node stakes 5% of the total amount of cryptocurrency staked, then, the node has a 5% chance of getting selected as a validator. Here, the reason behind selecting the nodes with the largest number of stakes is that the node is less likely to attack the network. However, the selection depending on the stake is not fair, as the richest node is more likely to dominate the network, which progressively resembles a centralized system [27].

3. **Practical Byzantine Fault Tolerance (PBFT):** PBFT is a consensus mechanism that allows the network to reach consensus even when some of the nodes in the network fail to respond or behave maliciously. The aim is to safeguard against system failures by reducing the influence of faulty nodes. PBFT requires all the nodes of the network to participate in the voting process, and at least, two-thirds of the total nodes agree to reach a consensus. This consensus algorithm has mainly three phases, namely pre-prepared, prepared, and commit. A block may be added to the blockchain, only if it receives a valid response in all the phases from two-thirds of the nodes. This algorithm can tolerate malicious activity from up to one-third of the total nodes, i.e., PBFT can function properly only if the maximum number of malicious nodes is less than one-third of the total nodes in the system. If there are (3m+1) number of correctly working processors, PBFT consensus can be reached, if at most m number of processors are faulty. This means strictly more than two-thirds of the total number of processors must be honest to reach PBFT consensus [28]. PBFT follows the following steps to validate a block:
   - The client sends requests to the server.
   - The server runs the three phases of the consensus process, i.e., pre-prepared, prepared, and commit.
   - The client receives the response whether it is validated or not.
4. **Proof-of-Activity (PoA):** PoA is another blockchain consensus algorithm that is developed by combining two consensus algorithms, namely PoW and PoS. PoA combines the best features of PoW and PoS systems, where blocks are mined through the PoW process, but once a new block is successfully mined, the system changes to look like a PoS system. Here, the newly mined block contains the miner's address and a header without any transaction. In PoA, based on the header details, a random group of validators is chosen for validating or signing the new block. A validator has more chances of being chosen as a signer, if s/he owns more coins. Like PoW and PoS, PoA also prevents 51% attacks. However, it needs too much power during block mining. Decred and Espers are two cryptocurrencies that use the PoA consensus algorithm.
5. **Stellar Consensus Protocol (SCP):** In 2015, David Mazieres first described SCP using Federated Byzantine Fault Tolerance (FBFT) [29]. This consensus algorithm is a decentralized and leaderless computing network that allows everyone to participate to reach a consensus efficiently. The Stellar payment network reaches consensus using SCP. In SCP, each participating core node selects a trusted set of other nodes to reach a consensus. This trusted set of nodes is called a quorum set. SCP supports robustness by using quorum sets. In

SCP, each consensus round has two steps: (i) nomination protocol and (ii) ballot protocol. The goal of the nomination protocol is to confirm the nomination of some values. Once a nominee is confirmed, it is called a candidate. Here, the candidate transaction sets are chosen for recording in a ledger. Each node gives a vote to select a unanimous single value among all the candidate values. The voting for new transaction set nominations stops once a node confirms its first candidate. As soon as a node confirms a candidate, the ballot protocol can be initiated. The ballot protocol makes sure that nominated transaction sets can be unanimously confirmed and applied on the network.

6. **Ripple:** Like SCP, ripple also uses the FBFT method, which was developed to reduce the latency. This decentralized method allows fast, low-cost, and decentralized transactions. In ripple, each miner uses a group of trusted nodes to reach a consensus. This consensus protocol has two types of nodes, namely server node and client node. Server nodes are responsible for consensus protocol and also for maintaining a list called Unique Node List (UNL), and the client nodes are responsible for the transfer of funds. If 80% of nodes in UNL agree on a transaction, then, the consensus is reached. This indicates ripple can tolerate only 20% of faulty nodes. However, the ripple network is ideal for financial transactions due to its overall low latency, speed, and security.

There are many other consensus algorithms, such as Proof of Elapsed Time (PoET), Delegated Proof of Stake (DPoS), Proof of Authority (PoA), Proof of Capacity (PoC), Proof of Proof, Proof of Burn, and Raft. A comparison between different consensus algorithms is shown in Table 1.3. However, rather than depending on a single consensus mechanism, nowadays, there is a trend of integrating a few consensus mechanisms to improve performance.

1.3.5 Smart contract

One of the key elements of a blockchain network is the smart contract. These are the self-executing rules and logic written in lines of code that a blockchain network

Table 1.3 Comparison among different consensus algorithms (CP = computing power, FR = faulty replicas)

Parameter	**Protocol**				
	PoW	**PoS**	**PBFT**	**DPoS**	**PoA**
Speed	Slow	Normal	Fast	Normal	Slow
Energy consumption	Very high	Normal	Very low	Normal	Normal
Security	Secure	Secure	Least secure	Secure	Secure
Blockchain type	Permissionless	Both	Both	Both	Permissioned
Scalability	High	High	Medium	Medium	High
Attack tolerance	< 25% of CP	< 51% of stake	< 33% of FR	< 51% of validators	< 25% of CP

Table 1.4 Comparison between traditional contract and smart contract

Parameters	Traditional contracts	Smart contracts
Execution time	1–3 days	Minutes
Remittance	Manual	Automatic
Security	Limited	Cryptographically secure
Cost	Expensive	Cheap
Signature	Physical signature	Digital signature
Lawyer	Necessary	May not be necessary
Data extraction	Slow and manually done	Immediate and automatic

uses for its terms, conditions, or automatically performed tasks. Smart contracts work on Boolean conditions, such as if-then-else. This feature was first introduced in the Ethereum platform [30]. All the aspects of Ethereum operations are governed by smart contracts. Smart contracts are the digital equivalent of contracts, which are triggered by any state change or transaction registration in the blockchain and are self-executing in nature. The smart contract has the ability to speed up commercial negotiation, validation, and execution without requiring a third party, and also provides direct contract execution between sender and receiver. In the Ethereum platform, these are written in Solidity programming language. Smart contracts are most useful in permissioned blockchain. It helps in completing any settlement fast. Smart contracts can automate many tasks of a blockchain network. Differences between traditional contracts and smart contracts are shown in Table 1.4.

1.3.5.1 Characteristics

In this subsection, some characteristics of smart contracts are discussed.

- *Independence:* It does not need any third party or mediator to authorize it as it is developed individually by a node.
- *Autonomous and decentralized:* As the codes in the smart contract are self-executing, it is autonomous. Furthermore, it cannot be modified or controlled by any centralized entity once it is deployed.
- *Trustfulness:* As these contracts are cryptographically encoded on a distributed ledger, no one can alter these contracts.
- *Transparent:* Anyone can see the smart contract's purpose and nature, when it executes on a public network.

1.3.5.2 Benefits

A few benefits of using smart contracts are listed below:

- *Reduced cost:* Smart contract reduces cost as it negates the existence of a third party.
- *Accurateness:* It reduces errors and processes faster and cheaper compared to traditional approaches.

- *Integrity:* Even if some nodes of the network exit the network, the smart contract continues to operate as per the pre-defined plan, which ensures network integrity.
- *Security:* As records are encrypted on a distributed ledger, it makes smart contracts very hard to hack.
- *Speed and efficiency:* As smart contracts are automated, the contract is executed immediately once a condition is satisfied. This saves time for business processes and also increases the efficiency of work.

1.3.5.3 Limitations

Along with the many benefits, smart contracts have some limitations, which can slow down the inner working of smart contracts execution. A few disadvantages of smart contacts that can limit their application in real-life scenarios are discussed below:

- *Built-in rules:* As smart contracts are rules and logic written in lines of code, there is a chance that the coder or programmer misinterprets and omits something that can result in contract flaws.
- *Difficulty to change:* Smart contract processes are very hard to change, and fixing any coding errors is time-consuming and expensive.
- *Exposure to bugs and errors:* Like regular codes, it can also develop bugs in it. However, solving bugs and errors in smart contracts is much costlier than regular codes and programs.
- *Legal settlement:* Smart contracts decrease the need for middlemen. Thus, all parties need to be conscious of the implications of public, private, criminal, and commercial law [31].

1.3.6 Hyperledger

This section provides a high-level overview of the Hyperledger foundation, which is the open and global ecosystem for improving blockchain technologies. Hyperledger was founded by Linux Foundation in 2015 to bring transparency and efficiency to Distributed Ledger Technologies (DLT) [32]. The aim of Hyperledger was to create an environment in which communities of developers may interact and work together to develop blockchain technologies. Hyperledger is not another cryptocurrency, but rather it can be considered a hub, where numerous independent blockchain-based projects and tools that follow its specified design tenet operate under its wing. Currently, eight projects are incubating under the Hyperledger Project. Five projects are based on the DLT, namely Hyperledger Iroha, Hyperledger Sawtooth, Hyperledger Fabric, Hyperledger Indy, and Hyperledger Burrow, and three projects are development tools that support those DLT frameworks, namely Hyperledger Cello, Hyperledger Explorer, and Hyperledger Composer. The overview of the various projects includes the following:

1. **Hyperledger Fabric** was first created by Digital Asset and IBM, and it is currently hosted by the Linux Foundation as a part of the Hyperledger project.

It is a blockchain development framework with a modular architecture for developing blockchain-based applications, solutions, and products for enterprises. The Hyperledger Fabric offers a high level of confidentiality, robustness, adaptability, and scalability as a private and permissioned blockchain. It has pluggable implementations of various components, including consensus and membership services among others. Hyperledger Fabric executes smart contracts in the form of programs and can be written in languages like Go, Python, java, etc. These smart contracts in the Fabric are called chaincode. Here, the only way to interact with the chaincode is through transactions.

2. **Hyperledger Iroha** is a blockchain framework, inspired by Hyperledger Fabric. It was created by the Israeli startup Colu, Hitachi, NTT Data, and the Japanese startup Soramitsu, and it is currently hosted by the Linux Foundation. Iroha provides distributed ledger technology to financial institutions and business organizations to manage digital assets. This provides libraries and components to integrate DLT into existing infrastructure. For example, Hyperledger Iroha can be used to deploy new currencies, bank-to-bank transfers, identity management, and many more.
3. **Hyperledger Sawtooth** is currently hosted by the Linux Foundation and was initially developed by Intel under its hyperledger project. It is a flexible and highly modular platform that can separate the core ledger system from the application domain. Thus, smart contracts can define the business rules for applications without being aware of the underlying architecture of the core system. This architecture enables programmers to develop applications in any programming language. Many consensus methods are supported by Hyperledger Sawtooth, including PBFT and PoET. It also allows running multiple consensus methods concurrently. PoET in Sawtooth has the advantage of low energy consumption with minimal resource utilization.
4. **Hyperledger Indy** is a Software Development Kit (SDK) that provides libraries, tools, and reusable components to generate globally unique digital identities on distributed ledgers. After creating a digital identity, users can give a name to their identities. On the distributed ledger, this identity name is changed into a special identity key called Decentralized Identifier (DID) that enables the communication between various people or organizations.
5. **Hyperledger Burrow** was initially developed by an open-source platform called Monax to efficiently execute smart contracts in a permissioned blockchain. This technology can create a complete, simple, and lightweight blockchain that can provide increased simplicity and speed to its users. Hyperledger Burrow fulfills the requirements of financial industries and investors by providing smart-contract-based applications. The main components of Hyperledger Burrow are the consensus engine, API gateway, smart contract application, Application Binary Interface (ABI), and Application Blockchain Interface (ABCI).
6. **Hyperledger Cello** is a framework that simplifies blockchain technology for normal users by acting as an intermediary layer between the blockchain and blockchain infrastructure. By using the cello operational dashboard,

blockchain developers can build, administer, and carry out other blockchain operations in a few easy steps. Hyperledger cello can quickly create a private blockchain and provide Blockchain-as-a-Service (BaaS) without doing complex programs. In Hyperledger Cello, basic characteristics, such as chain size, consensus technique, hosts, and many more, can be customized by the user.

7. **Hyperledger Composer** is a collection of collaborative tools for building and deploying blockchain into specific business areas. It uses the Hyperledger Fabric framework to validate transactions and enable protocols and policies. The modeling language provided by Hyperledger Composer enables users to define the assets, transactions, and participants, improving the business's effectiveness.
8. **Hyperledger Explorer** is an open-source tool to view the operations on a blockchain network. It can view, invoke, or query transactions, blocks, and data stored on the ledger. Hyperledger Explorer can easily display all information about a blockchain including its history, current status, and transactions on public networks. It can be considered as an interface for retrieving data stored on an open ledger.

1.4 Applications of blockchain technology

Even though blockchain technology was initially developed for solving issues in the financial sector, recently, it has expanded to other sectors like IoT, health care [33], and manufacturing [34], where it can improve transparency, manufacturing cost, and sustainability.

1.4.1 Applications of blockchain technology in healthcare

Conventional healthcare systems have been advanced over the years and adopted technological advancements to achieve better efficiency in the patient care system, as well as in the pharmaceutical industry [35]. As a result, the management of healthcare records and diagnosis of diseases have also improved. From related data management to tracking supply chain management of healthcare goods, blockchain technology is beneficial in almost every field of healthcare. Let us investigate the associated application areas, where blockchain can be used to create an effective e-healthcare system. A few application areas of blockchain technology in the healthcare domain are shown in Figure 1.7. This figure shows some of the key areas, where blockchain technology proves its effectiveness as a reliable solution.

Recently, many researchers suggested many techniques to use a blockchain network to store patients' data securely, which also supports the access to data of numerous patients without any privacy concerns [36].

1. *Health records management:* In healthcare, medical data are managed and stored as Electronic Health Records (EHR). EHR powered by blockchain technology can store information about a patient in an automatic manner on the blockchain network. As transactions on the blockchain are recorded immutably, EHRs cannot be altered by any hacker or attacker.

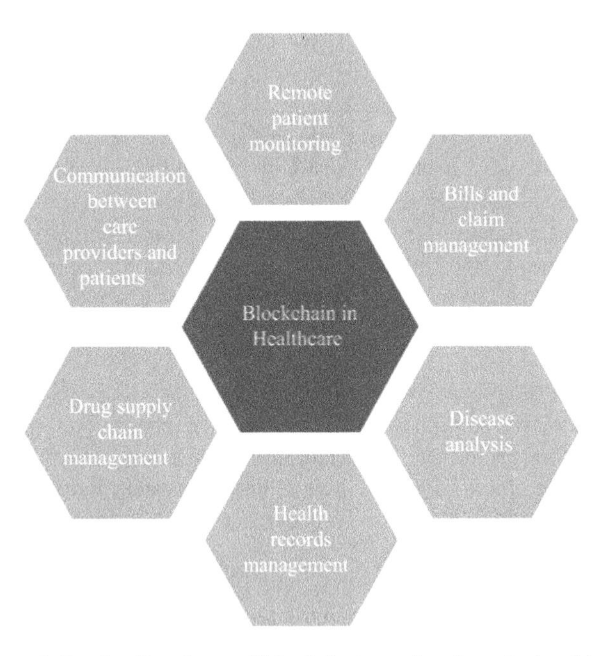

Figure 1.7 Applications of blockchain technology in healthcare

2. *Remote patient monitoring:* Patients' healthcare or medical data are very sensitive in nature and these data are usually stored in a centralized storage system for analysis. However, there exists a risk of data loss and tampering, when data are stored on a centralized system. Additionally, data transparency is also challenging to achieve, if all the data are stored in one location. By adopting a decentralized storage system, these data can be available to all users at any time. Here, blockchain solves all these issues by providing a secure and decentralized storage system.
3. *Drug supply chain management:* In the drug supply chain, blockchain manages inventory, and also, reduces forgery and theft issues. With the transparency, immutability, and auditability nature of blockchain, a blockchain-based drug supply chain management system enhances the supply chain's security, integrity, and functionality.
4. *Bills and claim management:* Several significant concerns in the healthcare sector, including manual errors, duplications, and improper billing, can be eliminated by using blockchain technology. The use of blockchain technology in the fields of claims settlement and billing administration would optimize the process of bill and claim settlement.

1.4.2 Applications of blockchain in banking and finance

Blockchain technology has many applications in the banking and finance sectors, such as payment systems, capital marketing, trading, loans and credit protection,

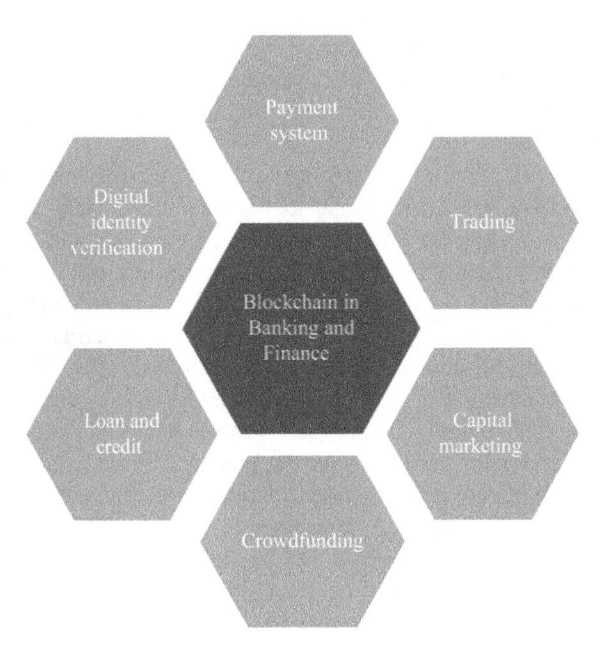

Figure 1.8 Applications of blockchain technology in banking and finance

crowdfunding, and many more. Figure 1.8 shows a few application areas of blockchain technology in the banking and finance sectors.

1. *Payment system:* In the banking and finance sector, faster payment and lower processing costs can improve customer satisfaction. By integrating blockchain technology, banks can diminish the need for third-party verification, which can speed up the payment process. Thus, both customers and banks can save time and money by using blockchain to execute financial transactions. One of the most well-known blockchain-based payment systems is called Ripple. The system enables direct money transfers between banks, businesses, and cryptocurrency exchanges without the use of a third-party mediator. For example, Mastercard presents patented blockchain technology that manages cryptocurrency transactions on conventional credit card networks.
2. *Capital marketing:* In capital markets, blockchain technology offers considerable benefits to issuers by enabling quicker, cheaper, and easier access to capital via programmable digital assets and securities. In 2013, a startup called Axoni develops a blockchain solution with a focus on the capital markets industry.
3. *Trading:* There are many participants in the traditional stock trading system, such as regulatory bodies, investors, brokers, and the centralized organization, for handling money. In the traditional stock trading system, an investment settlement takes several days to process due to the interaction among intermediaries. Blockchain technology has the potential to fundamentally change

the stock market by reducing costly, lengthy, and complex operations, as well as minimizes security issues. For example, Robinhood is one of the largest online trading platforms allowing investors to buy, sell, and trade cryptocurrencies.

4. *Loan and credit:* Traditional banking institutions assess loans using a credit reporting system. When processing loan applications, banks consider factors, including debt-to-income ratios, homeownership status, and credit ratings to determine risk. Due to this, the traditional processing can take many days to process the loan, and also, may face challenges like client identity verification, i.e., Know Your Customer (KYC), and money laundering laws, i.e., Bank Secrecy Act (BSA), and Anti-Money Laundering (AML). Financial services powered by blockchain can speed up and increase transparency in this process. The decentralized ledger of blockchain enables banks to share KYC, BSA, or AML processes with the customers by attaching these to a single customer block. Moreover, the blockchain-enabled system can streamline these processes by recording data and information on a decentralized ledger.
5. *Crowdfunding:* Crowdfunding is the practice of raising small amounts of money from a large number of people (hence the term "crowd") and the fundraising takes place via the internet [37]. Ata Plus is a blockchain-enabled licensed equity crowdfunding platform, which uses blockchain technology for keeping records of transactions [38].

1.4.3 Applications of blockchain in government

Most of the government and public sector institutions struggle to provide transparency, fairness, and accountability to the public. Blockchain-based distributed ledger technology can increase confidentiality and accountability while securing data, streamlining processes, and reducing fraud and misconduct. All government and public sector organizations can store data on a distributed ledger that is encrypted using cryptography. This design removes the risk of a single point of failure, as well as safeguards the privacy and confidentiality of information belonging to the public and government. Governments and public sector organizations can employ blockchain technology in the following situations [39]:

1. *Record management:* In every country, the government needs to hold the public identification, such as name, marital status, family details, birth date, property, and many more, of all its citizens. For example, in India, the government maintains PAN cards and Aadhaar information, which are used by many organizations, including banks and other financial institutions. However, the conventional data management system fails to update and verify these details in real-time. Additionally, citizens often need to go to local government offices physically for updating any changes in their documents. In such a scenario, a blockchain-based system can provide a secure means of managing and verifying this data and it can also enable citizens to jointly manage this data with the government. Figure 1.9 shows a few use cases of blockchain technology in the government sector.

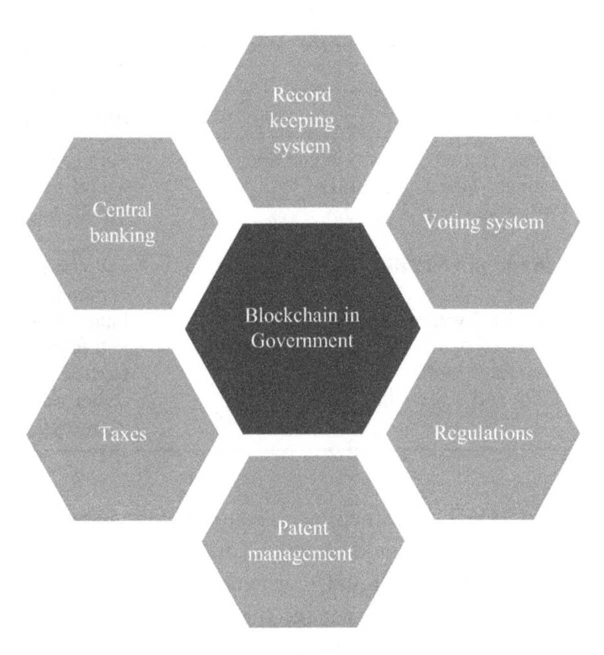

Figure 1.9 Applications of blockchain technology in government

2. *Voting:* There is no better technology than blockchain, which can ensure a transparent and highly secure voting process. Blockchain technology supports the voting process to be easy, transparent, and secure. The tamper-proof feature of blockchain technology prevents an attacker or malicious user from accessing the network server or system, thus, the attacker cannot tamper with other nodes of the blockchain network. Follow My Vote and Votem are examples of blockchain-based voting systems. The US government has employed a blockchain-based voting system in the 2020 presidential election and 2018 midterm election by using the commercial voting application Voatz [40].
3. *Patent management:* Government has a significant role in promoting technological advancement and industrial innovation. Effective patent management and registration can eliminate the theft of emerging technology and intellectual property. However, the existing patent authentication and management system are inefficient and slow. The immutability, trustworthiness, and security feature of blockchain can speed up the patent verification process, improve the patent registration and management process, and can solve any patent-related conflicts between parties [41].
4. *Taxes:* Manual tax filing may bring human errors in the tax filing process. Blockchain technology can make the process automatic and error-free by using smart contracts. Smart contracts automate the process by comparing tax information with income transactions, computing tax, and social security deductions. This automation makes tax collection more effective, swift, and secure.

Table 1.5 *Blockchain use cases adopted by countries and their aim*

Use cases	Country	Aim
Medical and healthcare	China, United States, Switzerland, Philippines, Japan, and Brazil	Supply chain management and IoT
Data management	Philippines and Australia	Cloud data management
Financial application	Almost all countries	Cryptocurrencies and asset management
Asset management	Georgia, Sweden, and Switzerland	Land registry and property transactions
Education	Japan and Malta	Certificate management
Government	Malta and Australia	Cyber security and data storage

5. *Central banking:* The government can create digital currency by using blockchain technology and provide seamless transaction facilities through the adoption of Central Bank Digital Currency (CBDC). In comparison to fiat currency, digital currency can manage the money supply and liquidity issues more efficiently. Moreover, blockchain can enable central banks to process Real Time Gross Settlement (RTGS) in a faster way with high security. Blockchain technology is adopted by many countries or governments to improve and automate several industries, which is shown in Table 1.5.

1.4.4 Applications of blockchain in manufacturing

Due to the varying customer expectations and evolving market dynamics, manufacturers face issues with demand forecasting, inventory control, managing manufacturing plant capacity, assuring ROI, and driving digital transformation. It is also challenging to track and trace the flow of raw materials, finished goods, and spare parts as they pass through production facilities, warehouses, distributors, and retail outlets due to complex and broad supply chains that are spread across large geographic areas. The following use cases show how blockchain might help manufacturers to overcome their problems in the manufacturing sector.

1. *Asset tracking and tracing:* Blockchain-enabled systems can track and trace every stage of the manufacturing process. When integrated with IoT devices, they can even aggregate, store, and share data with other parties across the supply chain. The IoT-enabled devices sense and collect data at every stage of manufacturing and store it on a secure and peer-to-peer network. As a result, all parties in a supply chain process can access all data at any time. Blockchain and IoT offer a more realistic solution with benefits, such as data integrity and transparency, as compared to conventional tracking techniques like Near-Field Communication (NFC), Global Positioning System (GPS), and Radio-Frequency Identification (RFID). Several use cases of blockchain technology in the manufacturing sector are shown in Figure 1.10.

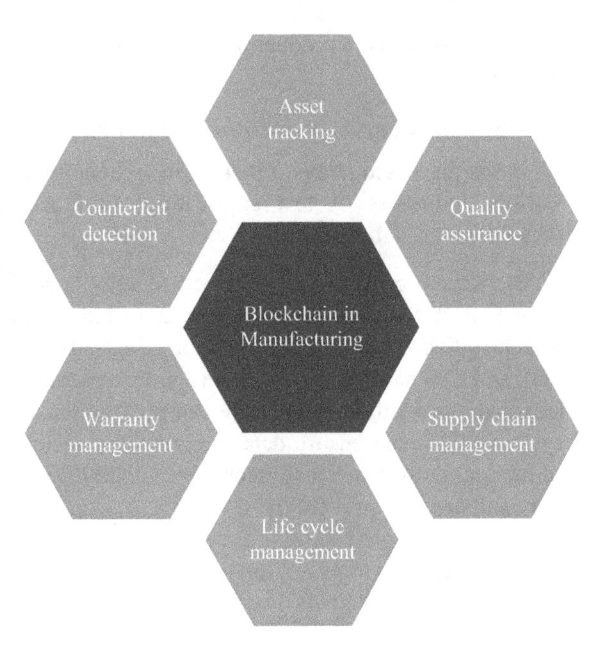

Figure 1.10 Applications of blockchain technology in manufacturing

2. *Warranty management:* In the manufacturing industry, proper warranty management can reduce false warranty claims, cut expenses, and provide an outstanding customer experience. There are numerous hurdles that businesses must overcome, ranging from false claims and fake products to misconceptions about coverage. Blockchain can solve all these hurdles as it provides an immutable record that cannot be altered or fabricated. Businesses can use blockchain technology to verify a product's information, including its serial number and ownership, and it can also verify genuine warranty claims. The advanced traceability approach provided by blockchain technology can be used to achieve excellence in warranty management.
3. *Quality assurance:* Quality assurance is another use case of blockchain in the manufacturing domain. Blockchain technology offers simplified and safe quality control checks by providing full transparency and immutable records. As individuals are involved in blockchain-enabled manufacturing systems, it can obtain complete transparency and the necessity for audits by central authorities is diminished.
4. *Product lifecycle management:* Manufacturers frequently lose track of where their products are being used by end users, when they sell their goods through integrators or distributors. Blockchain technology offers improved product lifecycle management. Here, it does not need complicated system integrations to establish a single data thread over the whole product life cycle. Additionally, the blockchain's timestamping feature and audit trail provide a precise record

of who, what, where, and when a product may have undergone changes during a specific period of time.

1.4.5 Applications of blockchain in other industries

Apart from the use cases discussed above, blockchain technology has many other applications in various industries. A few applications of blockchain in other industries are discussed in this subsection:

1. *Cybersecurity:* Blockchain can provide advanced security control by providing end-to-end encryption. A few well-known examples of blockchain-based cyber-security are Barclays traditional banking (London, England), CISCO IoT (San Jose, California), and Coinbase cryptocurrency (San Francisco, California).
2. *Big data:* Blockchain as a ledger technology can provide security and privacy in big data systems. This can ensure data integrity and accuracy in big data analytics. The integration of blockchain technology with big data can improve big data security and privacy, prevent fraud activity, enhance the quality of big data and data sharing, and automate data access techniques.
3. *IoT:* Currently, trade freight logistics use blockchain to automate IoT processes. There are many stakeholders involved in freight logistics, including manufacturers, insurers, custom agents, and shippers, who frequently interact with each other to achieve different goals and employ various tracking systems to track goods. IoT-enabled blockchain is used as a shared ledger to record shipping containers as they pass through the system. Here, IoT or sensor devices use smart contracts to track and automatically update the tracking status of goods in IoT-enabled blockchain network.
4. *Agriculture industry:* Blockchain has many use cases in the agriculture industry. Blockchain-enabled agricultural systems can track all information about plants, including the quality of the seed and crop growth. This can even track the plant's journey after leaving the farm through the agricultural supply chain process. This can improve food safety and reduce illegal and unethical production issues by increasing the transparency of supply chains. Moreover, fraud and malfunctions can be detected in real-time by incorporating smart contracts.

1.5 Challenges in adopting blockchain technology

Blockchain technology has much hypes, which has made many organizations in different industry sectors eager to adopt it. However, as this technology evolves, it brings many issues, as well as dependencies on other technologies. A few issues with the adoption of blockchain technology are listed below:

1. *High energy consumption:* Many businesses avoid using blockchain technology because of the huge amount of energy required to run a blockchain-based application. For example, both Ethereum and Bitcoin networks use PoW as a consensus algorithm to validate transactions, which requires the computation of complex mathematical problems. Solving these problems requires a large amount of energy to power the computer nodes on the network.

2. *Inefficient technological design:* The technological design of blockchain is another major challenge. Although Bitcoin was the pioneer in this area, the system has a poor system design. Ethereum made an effort to address all of Bitcoin's shortcomings, but, it was insufficient. A coding flaw or loophole is one of the significant points in this.
3. *Lack of flexibility:* Due to the immutable nature of blockchain, a state cannot be changed once it is created. This feature can be considered a double-edged sword. As the data cannot be changed after they are recorded, the trust regarding data integrity is increased, and also, the risk of fraud is reduced. However, this feature eliminates the ability to modify incorrect records for valid reasons. Therefore, the lack of flexibility in data modification is considered a challenge, when adopting blockchain in many sectors.
4. *Regulatory challenges:* Apart from technological issues, the lack of regulation clarity also makes the adoption of blockchain technology challenging. Current regulatory frameworks cannot sustain the rapid development of blockchain-based technology. As there are no explicit rules governing it, nobody strictly adheres to any regulations, when it comes to the blockchain.
5. *Lack of expertise:* There are very less people with the expertise to support blockchain technology. To design a blockchain-based solution, professionals from organizations are required to adopt this technology. Additionally, specialized knowledge is needed to design a secure and reliable blockchain-based system. The lack of appropriate training and awareness about the benefits of blockchain has made the adoption of this technology challenging.
6. *Financial barriers:* Blockchain-based system is costlier than traditional systems. The transaction costs also cannot be disregarded while adopting blockchain technology. Moreover, there are infrastructure costs, computational energy costs, and time costs due to slow transactions in a blockchain network. Furthermore, the addition of new features can result in high costs each time.

1.6 Conclusions

Blockchain is an immutable public ledger, which records the data shared among all the participants of the network. Blockchain technology allows tracing an asset's ownership using the distributed ledger, which makes it difficult for cybercriminals and hackers to manipulate data. However, it can be difficult for users to comprehend the technologies it provides because of the complex structure. To address this, in this chapter, the basics of blockchain technology are covered, including its history, characteristics, benefits, drawbacks, and many other things. The technological aspects of blockchain are also explained in detail. Several use cases of blockchain technology in many fields, such as government and public sector, banking and finance, healthcare, and manufacturing sector, are also discussed in this chapter. Moreover, challenges in adopting blockchain technology in several sectors are also represented in detail. This could be helpful for researchers, academics, and business experts to adopt blockchain in businesses to improve the work process of traditional systems.

References

[1] S. Jangirala, A.K. Das, and A.V. Vasilakos, "Designing secure lightweight blockchain-enabled RFID-based authentication protocol for supply chains in 5G mobile edge computing environment", *IEEE Transactions on Industrial Informatics*, vol. 16, no. 11, pp. 7081–7093, 2020.

[2] T.M. Fernndez-Carams and P. Fraga-Lamas, "A review on the use of blockchain for the Internet of things", *IEEE Access*, vol. 6, pp. 32979–33001, 2018.

[3] S. Namasudra, G.C. Deka, P. Johri, M. Hosseinpour, and A.H. Gandomi, "The revolution of blockchain: state-of-the-art and research challenges", *Archives of Computational Methods in Engineering*, vol. 28, pp. 1497–1515, 2021.

[4] Y. Hu, A. Manzoor, P. Ekparinya, *et al.*, "A delay-tolerant payment scheme based on the Ethereum Blockchain", *IEEE Access*, vol. 7, pp. 33159–33172, 2019.

[5] I. Miers, C. Garman, M. Green, and A.D. Rubin, "Zerocoin: anonymous distributed E-cash from Bitcoin", In *Proceedings of the IEEE Symposium on Security Privacy*, IEEE, 2013, pp. 397–411.

[6] E.B. Sasson, A. Chiesa, C. Garman, *et al.*, "Zerocash: decentralized anonymous payments from Bitcoin", In *Proceedings of the IEEE Symposium on Security and Privacy*, IEEE, Berkeley, CA, 2014, pp. 459–474.

[7] J. Sidhu, "Syscoin: a peer-to-peer electronic cash system with Blockchain based services for E-business", In *Proceedings of the 26th International Conference on Computer Communication and Networks (ICCCN)*, IEEE, Vancouver, BC, Canada, 2017, pp. 1–6.

[8] S. Namasudra and P. Sharma, "Achieving a decentralized and secure cab sharing system using blockchain technology", *IEEE Transactions on Intelligent Transportation Systems*, 2022, doi: 10.1109/TITS.2022.3186361.

[9] A. Dorri, S.S. Kanher, R. Jurdak, and P. Gauravaram, LSB: a Lightweight Scalable Blockchain for IoT Security and Privacy. Available: https://arxiv.org/abs/1712.02969?context=cs, 2017 [Accessed on 9 Sept 2022].

[10] Y. Zhang, S. Kasahara, Y. Shen, X. Jiang, and J. Wan, Smart contract-Based Access Control for the Internet of Things. Available: https://arxiv.org/abs/1802.04410, 2018 [Accessed on 9 Sept 2022].

[11] H. Kopp, D. Modinger, F. Hauck, F. Kargl, and C. Bosch, "Design of a privacy-preserving decentralized file storage with financial incentives", In *Proceedings of the IEEE European Symposium on Security and Privacy Workshops*, IEEE, Paris, France, 2017, pp 14–22.

[12] H. Kopp, C. Bosch, F. Kargl, "KopperCoin—a distributed file storage with financial incentives", In *Proceedings of the International Conference on Information Security Practice and Experience*, New York, NY: Springer, 2016, pp. 79–93.

[13] X. Liang, S. Shetty, D. Tosh, C. Kamhoua, K. Kwiat, and L. Njilla, "ProvChain: a blockchain-based data provenance architecture in cloud environment with enhanced privacy and availability", In *Proceedings of the*

17th IEEE/ACM International Symposium on Cluster, Cloud and Grid Computing (CCGRID), Madrid, Spain, 2017, pp. 468–477.

[14] Q.I. Xia, E.B. Sifah, K.O. Asamoah, J. Gao, X. Du, and M. Guizani, "MeDShare: trust-less medical data sharing among cloud service providers via blockchain", *IEEE Access*, vol. 5, pp. 14757–14767, 2017.

[15] P. Dunphy and F.A.P. Petitcolas, A First Look at Identity Management Schemes on the Blockchain. Available: https://arxiv.org/abs/1801.03294, 2018 [Accessed on 11 Sept 2022].

[16] S. Raju, S. Boddepalli, S. Gampa, Q. Yan, and J.S. Deogun, "Identity management using blockchain for cognitive cellular networks", In *Proceedings of the IEEE International Conference on Communications*, IEEE, Paris, France, 2017, pp. 1–6.

[17] S. Muftic, Blockchain Identity Management System based on Public Identities Ledger, 2017. U.S. Patent 9635000.

[18] A.T. Sherman, F. Javani, H. Zhang and E. Golaszewski, "On the origins and variations of blockchain technologies", *IEEE Security & Privacy*, vol. 17, no. 1, pp. 72–77, 2019.

[19] D. Chaum, "Computer systems established, maintained, and trusted by mutually suspicious groups", Dissertation, Computer Science, University of California, Berkeley, *CA*, 1982.

[20] A. Narayanan, J. Bonneau, E. Felten, A. Miller, and S. Goldfeder, *Bitcoin and Cryptocurrency Technologies: A Comprehensive Introduction*, Princeton, NJ: Princeton University Press, 2016, ISBN 978-0-691-17169-2.

[21] S. Haber and W.S. Stornetta, "How to time-stamp a digital document", *Journal of Cryptology*, vol. 3, no. 2, pp. 99–111, 1991.

[22] D. Bayer, S. Haber, and W.S. Stornetta, "Improving the efficiency and reliability of digital time-stamping", *Sequences*, vol. 2. pp. 329–334, 1992.

[23] M.E. Peck, "The cryptoanarchists' answer to cash", *IEEE Spectrum*, vol. 49, no. 6, pp. 50–56, 2012.

[24] S. Nakamoto, Bitcoin: A Peer-To-Peer Electronic Cash System. Available: https://bitcoin.org/bitcoin.pdf, 2008 [Accessed on 16 Sept 2022].

[25] J. Golosova and A. Romanovs, "The advantages and disadvantages of the blockchain technology," In *2018 IEEE 6th Workshop on Advances in Information, Electronic and Electrical Engineering (AIEEE)*, 2018, pp. 1–6, doi:10.1109/AIEEE.2018.8592253

[26] G. Pirlea and I. Sergey, "Mechanising blockchain consensus", In *Proceedings of the 7th ACM SIGPLAN International Conference on Certified Programs and Proofs*, ACM, 2018, pp. 78–90.

[27] S. King and S. Nadal, Ppcoin: Peer-to-peer Crypto-Currency with Proof-of-Stake. Available: https://peercoin.net/960assets/paper/peercoin-paper.pdf, 2012 [Accessed on 11 Sept 2022].

[28] M. Castro and B. Liskov, "Practical byzantine fault tolerance", In *Proceedings of the T3rd Symposium on Operating Systems Design and Implementation (OSDI'99)*, USENIX Association, Berkeley, CA, 1999, pp. 173–186. http://dl.acm.org/citation.cfm?id=296806.296824

[29] S. Barański, J. Szymański, A. Sobecki, D. Gil, and H. Mora, "Practical I – Voting on stellar blockchain", *Applied Sciences*, vol. 10, no. 21, pp. 1–22, 2020, doi: 10.3390/app10217606
[30] V. Buterin, A Next-Generation Smart Contract and Decentralized Application Platform. Available: https://blockchainlab.com/pdf/Ethereum_white_paper-a_next_generation_smart_contract_and_decentralized_application_platform-vitalik-buterin.pdf, 2014 [Accessed on 11 Sept 2022].
[31] H.M. Kim and M. Laskowski, "Toward an ontology-driven blockchain design for supply-chain provenance", *Intelligent Systems in Accounting, Finance, and Management*, vol. 25, no. 1, pp. 18–27, 2018.
[32] A.S. Gillis, Hyperledger. Available: https://www.techtarget.com/searchcio/definition/Hyperledger#:~:text=Hyperledger%20is%20an%20open%20source, build%20blockchains%20and%20related%20applications. [Accessed on 11 Sept 2022].
[33] A. Firdaus, M.F.A. Razak, A. Feizollah, I.A.T. Hashem, M. Hazim, and N.B. Anuar, "The rise of 'blockchain': bibliometric analysis of blockchain study", *Scientometrics*, vol. 120, no. 3, pp. 1289–1331, 2019.
[34] T. Ko, J. Lee, and D. Ryu, "Blockchain technology and manufacturing industry: real-time transparency and cost savings", *Sustainability*, vol. 10, no. 11, pp. 1–20, 2018.
[35] S. Das and S. Namasudra, "Multi-authority CP-ABE-based access control model for IoT-enabled healthcare infrastructure", *IEEE Transactions on Industrial Informatics*, vol. 19, no. 1, pp. 821–829, 2022, doi: 10.1109/TII.2022.
[36] S. Namasudra and G.C. Deka, *Applications of Blockchain in Healthcare*, New York, NY: Springer, 2021, doi: 10.1007/978-981-15-9547-9.
[37] The Application of Blockchain Technology in Crowdfunding: Towards Financial Inclusion via Technology. Available: https://www.researchgate.net/publication/327586306_The_Application_of_Blockchain_Technology_in_Crowdfunding_Towards_Financial_Inclusion_via_Technology [Accessed on 11 Sept 2022].
[38] K.A. Noordin, Profile: Putting Her Faith in Equity Crowdfunding. Available: http://www.theedgemarkets.com/article/profile-putting-her-faith-equity-crowdfunding, 2018, [Accessed on 11 Sept 2022].
[39] Blockchain in Government and the Public Sector. Available: https://consensys.net/blockchain-use-cases/government-and-the-public-sector/ [Accessed on 11 Sept 2022].
[40] Which Countries are Casting Votes Using Blockchain? Available: https://hackernoon.com/which-countries-are-casting-voting-using-blockchain-s33j34ab [Accessed on 6 Sept 2022].
[41] P.T. Gunasekara and C. Rajapakse, "A blockchain-based model to improve patent authentication and management process", In *Proceedings of International Conference on Advanced Research in Computing (ICARC)*, 2022, pp. 338–343.

Chapter 2

e-Healthcare record management using blockchain technology

Shipra Swati[1] and Mukesh Kumar[1]

Abstract

Electronic health records (EHR) provide an end-to-end ecosystem that unites patients, healthcare providers, and pharmacists. For the sustainable and enhanced performance of the healthcare industry of any province, EHR has become a key requirement. If a secure storage and transmission mechanism are incorporated into the infrastructure, it has the capability to support architecture based on Internet-of-Things (IoT). Such distributed settings may be equipped with blockchain technology to guard sensitive information regarding a person's medical status. Blockchain has been emerged as the most trustworthy technology for mitigating the risk of cyber-attacks and frauds in banking, cryptocurrency, insurance, supply chain, and healthcare as well. This chapter shows how blockchain technology can be used to monitor and manage patient data, prescriptions, and medicine traceability with minimum or no falsification of real data. A five-layer architecture has been presented in this study using Blockchain-based IoT (BIoT) for decentralized traceability in the medicine supply chain. The viability and effectiveness of BIoT have been validated by utilizing the real data from the Ethereum Blockchain ledger. The pharmaceutical industry can also utilize the proposed framework to implement blockchain design for robust supervision of the drug-supply chain. The amalgamation of transparency in information-sharing among collaborators will also help in regulating the quality of medicine.

Keywords: EHRs; Healthcare; Patient data; Prescription; Drug traceability

2.1 Introduction

Electronic health records (EHRs) provide persistent support for clinical practice and clinical assessment. The incorporated execution indicators are generally defined as longitudinal data (in an electronic course of action) that are gathered

[1]Department of Computer Science and Engineering, National Institute of Technology Patna, India

during routine clinical care transit [1,2]. EHRs typically comprise critical estimations, administrative, clinical, and patient-centered data (gathered through various modes, such as individuals with instruments or equipment to understand the behavior of the human body, home-noticing gadgets, and private blockchain networks, like the Gatekeeper). Concerns have been raised about the rising selection challenges in primers, harsh and unmistakable data variety, and uncertain generalizability of the results. The use of Internet of Things (IoT)-based devices have profoundly changed the healthcare sector [3]. IoT-based medical equipment and devices can assist in the remote monitoring of patient-health, and give patients more control over their lives and prescriptions [4–6]. The use of IoT devices in the healthcare sector has rapidly risen, and there are now a large number of smart wearable devices being used to measure things like blood pressure, blood sugar, oxygen saturation, and heart rate. These wearable devices are used to monitor, update, and maintain the patient's real-time data in their respective EHRs in addition to being able to make the diagnosis. In such a distributed environment, EHR is based on the idea of peer-to-peer (P2P) networking. An EHR has all the important information about a person's health care on the patient's side, and it also has all the information about the healthcare professionals and organizations that help people with their health care [7]. It also has all the information about pharmaceuticals, like the supply chain, drug details, prescription management, transactions, etc. All of this information is very sensitive and needs a system that is both safe to store and safe to use when information exchange is taking place. EHR exchanges can be made simple and secure using a blockchain organization, whose fundamental characteristic is the security of transactions.

Blockchain technology has already been used in a lot of different fields including healthcare, like banking, supply chain, energy, commodities trading, and a lot of businesses that process transactions [8,9]. The technology behind blockchain makes it possible to create a distributed shared data platform, which can then be used to store and share transaction information among the various supply chain stakeholders. It is a shared, decentralized database that keeps track of transactions in a growing chain of unchangeable blocks linked by cryptographic hashes. Blockchain is a better choice for monitoring and managing EHR data in a safe way because of its key features, such as transparency, authenticity, unchangeability, flexibility, and safe. Figure 2.1 shows how the blockchain and healthcare system are related. Additionally, this technology has the potential to provide a secure and unchangeable solution for drug traceability and provenance, which would be helpful in combating the problem of fake pharmaceuticals.

Moving ahead to the next step let's explore what are the related applications domains where the blockchain is applicable for an efficient e-healthcare system. Figure 2.2 shows the various application domains of blockchain technology in managing and monitoring e-healthcare data. Using blockchain technology is beneficial in almost every aspect of healthcare, from related data management to tacking supply chain management of healthcare goods. The figure shows some of the key areas where blockchain technology proves to be a reliable solution. The vital advantages of applying blockchain innovation in medical services are the

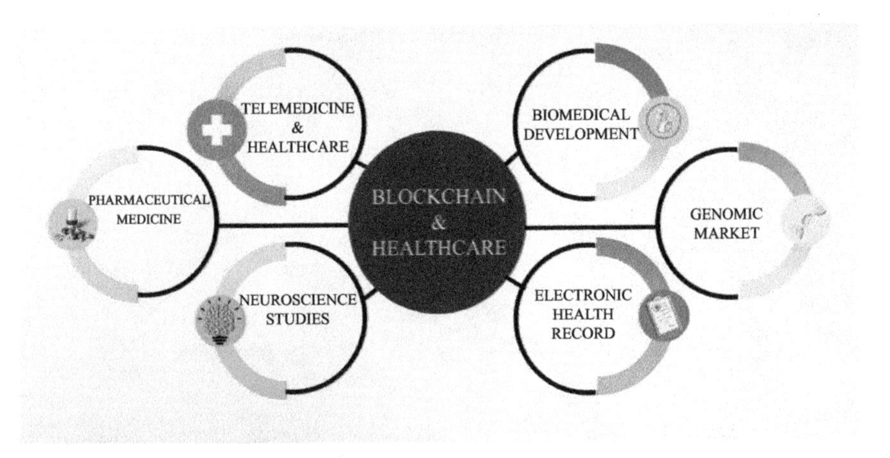

Figure 2.1 Relationship between healthcare and blockchain

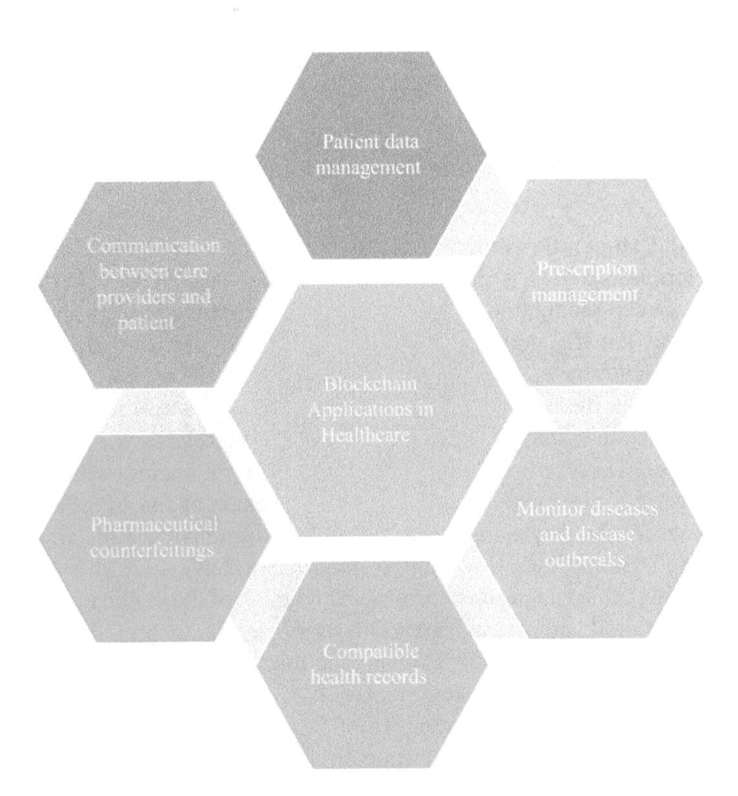

Figure 2.2 Applications of blockchain technology in health care

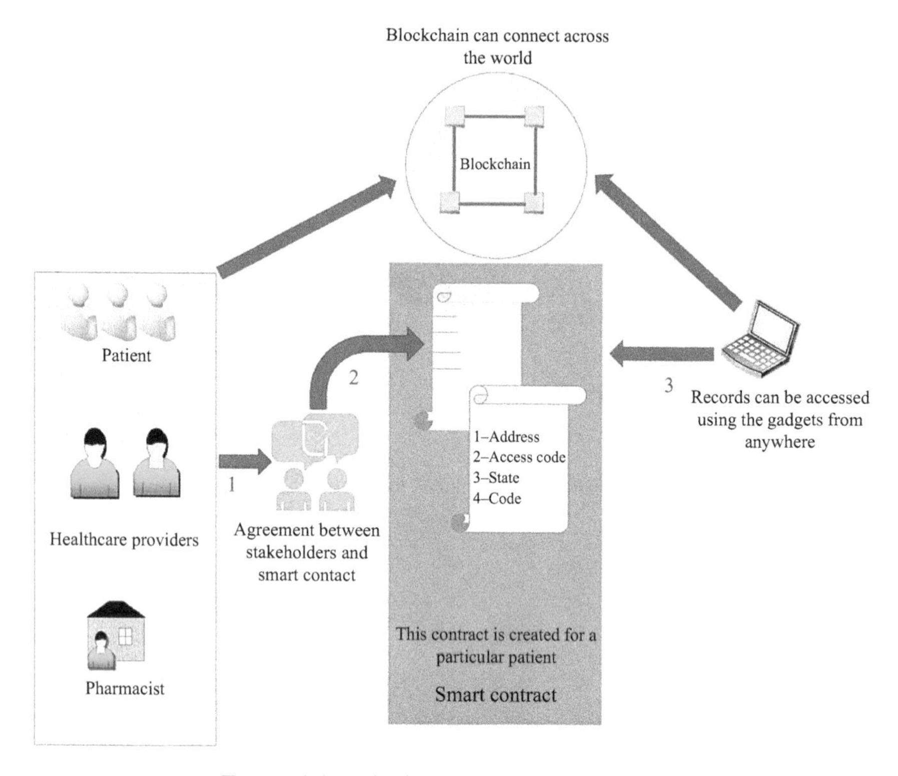

Figure 2.3 Blockchain in medical industry

accompanying straightforwardness, and trustworthiness of appropriated delicate clinical information. This is predominantly accomplished by utilizing agreement convention and cryptographic natives, for example, hashing and advanced marks.

There was not the slightest lingering doubt in anyone's mind that a decentralized application based on blockchain could be established for each and every point of contact in the health system. It is easy to understand from Figure 2.3 that how the production of data and its subsequent exchange feeds the ecosystem that supports the delivery of healthcare. This ecosystem begins with biomedical research carried out in a research lab with both the examination of cells and tissues and continues all the way through the payment of insurance claims after care has been provided. But on the other hand, in sequence to have a health system that is fully operational, the data needs to be distributed to all of the vital elements in order to ensure continuity in patient care. This is necessary to provide a healthcare delivery system that is fully operational. The mechanism of sharing the relevant data across the ecological system makes blockchain technology a beneficial implementation in the medical sector [16].

Several features of blockchain technology lend themselves to one steady basis, which is the capability to equitably negotiate the suspense between both data sharing as well as privacy. This is an undisputable fact about this technique and it is unanimously accepted by the research community. Over the span of decades, healthcare delivery organizations, pharmaceutical companies, medical practices, as well as other providers of healthcare services depend on policy in order to sustain the compartmentalization and security of a valuable asset, such as patient data. These organizations not only getting benefitted from the rewards of trying to leverage the data, but they are also trying hard to promise a healthy balance between the communication of patient information and the safeguards of their confidentiality.

Recently, there has been a lot of interest in how blockchains can be used to send completely secure medical services information [10,11], share biomedical [12], and e-health information [13], and help with brain exercise and thinking. However, a complete solution is required for handling multi-field structure and various numerical articulations. The main contributions of this presented work are as follows:

- This study proposes a distributed traceability solution for the management of healthcare records that uses IoT and Blockchain technologies to make sure that healthcare data can be tracked in a smart way using Blockchain-based IoT (BIoT).
- A five-layer blockchain platform architecture is suggested by authors to provide a complete solution for the drug-related supply-chain management system.
- The utility of machine learning in conjunction with blockchain technology is investigated for incorporating drug recommendations by the proposed framework.

This chapter is further organized as: Section 2.2 explains about electronic health records, Section 2.3 includes monitoring and managing e-healthcare data using blockchain technology, here Section 2.3.1 illustrates the patient data management, Section 2.3.2 includes prescription management, Section 2.4 includes drug traceability using blockchain technology, Section 2.5 discusses some important use cases. After a brief overview of related challenges in Section 2.6, the chapter concludes in Section 2.7.

2.2 Electronic health record

One may think of EHRs as a digitized version of a person's personal health data. The patient's financial history is combined in the EHR for safe organization and access in order to provide reliable care. EHRs were initially developed and used in academic clinical offices, but they are replaced by large commercial EHRs now. Their operation begins by including the tasks necessary to create any clinical record, such as analyzing essential information, which includes the analysis of the patients' significant bodily functions and socioeconomic status, their current medications and sensitivity levels, their current smoking status, and state-of-the-art issue for arranging current and dynamic judgments. Utilizing a few programming tools that would comprehend the true potential of EHRs by improving the security, quality, and proficiency of patient care is one of the major ambitions. It may aid clinicians in making better therapeutic decisions along with handling avoidable blunders.

EHRs may also be combined with data analysis for evaluation and advancement of effective pharmaceuticals and therapies for recurrent disorders. They are termed "Absolute level EHR" [14]. Data admission is computerized with the EHR, which may further facilitate the clinician's workflow. Additionally, other consideration-related activities, including evidence-based decision support, quality management, and outcomes proclaiming, can be supported directly or indirectly by the EHR through various connection points. EHRs and the ability to exchange health data electronically can help with providing patients with better and more secure care while also significantly enhancing the association with health professionals. Thus, EHRs can be concluded as a common architectural platform serving patients, doctors, and hospitals equally as shown in Figure 2.4.

2.2.1 Benefits of using EHRs

EHRs improve patient care and automate practice chores. They also allow physicians to exchange information remotely and in real-time, ensuring every clinician treating a patient gets a complete, correct file. They are also customized according to the requirement of the medical practitioner. Significant advantages of using EHR are listed below:

1. ***Improved care:*** EHRs allow faster access to patient records for efficient care. They improve therapeutic efficacy and practice efficiency. Most EHRs provide doctors with health analytics to identify patterns, diagnose, and suggest treatments. Instead of trial-and-error, these algorithms produce better patient outcomes. Patients can obtain lab and imaging data, prescriptions, diagnoses, and more through patient portals. They can also communicate with doctors through sharing notes, quick chats, and video calls. The platform lets doctors and patients follow therapy progress that have a positive impact on preventive care.
2. ***Better patient data:*** Electronic medical records have several benefits beyond making patient data easier to store and retrieve. Electronic storage eliminates

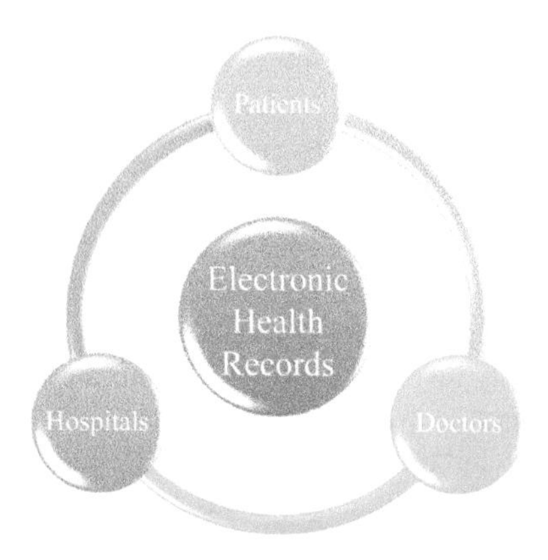

Figure 2.4 E-Health records architectural platform

the risk of data theft, loss, damage, or change. Digital records eliminate handwriting and readability mistakes. Physicians may update patient information in real-time, giving other healthcare providers updated patient files. This computerized record connects all clinicians and specialists involved in a patient's care. Continuity is important when a patient switches providers or visits a new doctor since it gives them a complete medical history. EHRs improve patient data accuracy by organizing information properly.

3. ***Effectiveness:*** EHRs help doctors diagnose and treat more accurately and save time. They expedite appointments and office visits without compromising patient care, allowing doctors to see more patients every day. EHR templates let doctors document frequent patient grievances. These templates are customized for specific professions or physicians through the support of AI. Most hospitals and specialties use e-prescribing that eliminates dependency on their location. Further, prescriptions can be forwarded electronically to the pharmacist. Based on a patient's prescription and diagnosis, the EHR can automatically check for drug-to-drug or drug-to-allergy interactions.
4. ***Revenue growth:*** Every business wants to maximize revenue including healthcare organizations. EHR tools manage revenue and payment by minimizing mistakes and coding errors from claims, preventing rejections. This function boosts acceptance rates for first-time insurance claims and speed up reimbursements without delay. EHR let doctors to document every detail of a patient's visit, making it easy to prove claims. EHR users receive government subsidies. EHR software helps medical practices follow rules and prepare for regulations. Government incentives make EHR implementation affordable.

Apart from the above-mentioned benefits, EHR also supports security in the scalable environment for providing ease of accessibility. Data migration, updates, and patches are usually handled automatically.

2.2.2 Risks of EHR

The idea behind the introduction of EHR programs was to address the issues with paper-based medical records and to include a useful tool that would improve the state of affairs in the healthcare sector. Many hospitals around the world are implementing EHR systems because of the benefits it provides, particularly the improvement of security and cost-effectiveness. They are regarded as an essential component of the healthcare industry because they supply medical institutions with a range of functionality. These capabilities are readily available in the majority of sectors of the EHR scheme in the healthcare industry, including the storing of electronic medical records, scheduling of patient appointments, billing and accounts, and laboratory testing. The main goal is to deliver secure, tamperproof medical information via various media. Even though the goal of the development of EHR healthcare systems was to increase efficiency, these systems encountered a number of issues and fell short of the requirements and standards that went along with them, such as being unreliable and in need of repair. When utilizing the EHR technique, the following problems can arise:

1. ***Interoperability:*** This is the ability of various information systems to exchange information with one another. Information ought to be able to be shared and used for extra reasons. The health information exchange (HIE) or, more generally, the sharing of data, is a crucial component of EHR systems. Furthermore, technical interpretation of the shared medical documents is necessary; this knowledge may then be applied.
2. ***Information asymmetry:*** According to opponents, the main issue facing the healthcare industry today is knowledge asymmetry, which refers to one side having greater access to information than the other. This problem arises when doctors or hospitals need to access the patient's records when implementing EHR schemes. When a patient desires to obtain his own medical records, extremely drawn-out and frustrating procedures must be followed in order to do so. Only one healthcare organization has central storage for the data, and relatively few hospitals or organizations have access to it.
3. ***Data breaches:*** Data breaches in the healthcare industry also necessitate the need for a more powerful forum. Since October 2009, the security of EHR systems has been breached, compromising about 173 million data entries. Many EHR systems also have issues with efficiency and poor adaption because they are not set up to suit patient demands and expectations. Additionally, according to the research, using EHRs negatively affects how information is processed. Lifelong, multi-institutional medical records were not intended to be supported by EHRs. As a patient's life events move them from the data silo of one provider and into another, data is spread out across many organizations. Due to the fact that primary stewardship is typically held by the clinician rather than the patient, they lose simple access to prior data in doing so. Figure 2.5 reveals the major benefits and risks of EHR for providing a simple overview.

By constructively establishing the specifications and standards for interoperability, which address privacy and allow for the secure sharing of data between systems, the blockchain architecture will undoubtedly aid in the resolution of this issue. Interoperability and open standards play a crucial part in enhancing the framework's flexibility and facilitating the exchange of health data. Due to these problems, it makes sense to look for the most efficient way to revolutionize the patient-centered healthcare sector, which is blockchain. A data integrity portal also

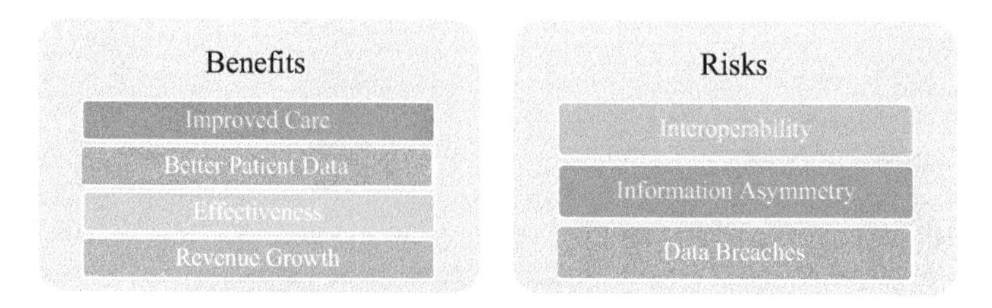

Figure 2.5 Benefits and risks of EHR

offers the patients' medical histories while being safe and open. Health records are estimations of a person's health through time, and a steady population is necessary to achieve the desired improvement. The entire globe is making development. Data breaches occur as a result of the centralized management of health records. Therefore, there was no difference between the old record management technique and the one that was instituted. The patient has no control over the data, which increases the likelihood that it will be misused. Therefore, we require a truly decentralized, patient-centered solution that has the ability to identify data thefts, prevent data abuse, and grant patients the right to access control. The best method for resolving all problems and fulfilling healthcare requirements is blockchain technology. It will have an impact on billing, record sharing, medical testing, fraud identification, and financial data crimes in the future as a decentralized and distributed ledger. Things will become much more straightforward if smart healthcare contracts are implemented. During the invocation, record creation, and record validation, it will be done utilizing blockchain.

2.3 Monitoring and managing e-healthcare data using blockchain technology

EHR for the most part contains exceptionally delicate and basic information connected with patients, which is frequently distributed among clinicians, radiologists, medical care suppliers, drug specialists, and scientists, for powerful analysis and therapy. During the capacity, transmission, and conveyance of this exceptionally delicate patient data among a few elements, the patient's treatment can be compromised, which can present extreme dangers to the patient well-being. On account of patients battling with persistent sicknesses (e.g., malignant growth and HIV), the pervasiveness of such dangers can become higher because of a long history of pre-and-post therapy, subsequent meet-ups, and recovery processes. In this way, having a forward-thinking for keeping patient history has become exceptionally basic in order to guarantee viable treatment. Different difficulties are faced by the medical care industry in digitizing and sharing clinical records, tracking follow-up of physician-recommended drugs and other clinical necessities, and storing and administering them over the network. To evade such limits, Dubovitskaya *et al.* [15] proposed a blockchain-based system for dealing with these issues related to the electronic clinical records of patients. They embraced a permissioned blockchain innovation to store encoded patient information. Such proposed structures can be utilized to execute blockchain innovation for getting to and dealing with the protection and security of patient information and history in clinical practices.

The above discussion justifies the major question, "what are the requirements of an EHR, and what we get in turn while using blockchain technology." Table 2.1 mentions and helps to understand this relationship of required vs. promised between EHR and blockchain.

Managing E-health care data requires two major data repositories to work well, namely patient data/history and prescription data. Patient data mainly consists of

Table 2.1 EHR requirements vs. blockchain promises

EHR requirements	Blockchain promises
Private and confidential	Security and authenticity
May not be amended	Indestructible
Audit trail database	Keeps history records
Sharing among multiple actors	Dispersed data sharing
Immediate turnaround and exchange	Efficacious

the patient's personal information along with the diagnosis, findings, corresponding treatments, and other medical procedures. All these records are mandatory to find the exact condition and health status of the patient and for proper handling by doctors/medical care persons. Another important aspect is to have a complete record of the patient's prescriptions and medicines prescribed to him for a particular treatment scenario. On the basis of these prescriptions' healthcare persons can easily find suitable medicines and also allergic salts and side effects if the patient faces some conditions. The following sub-sections discuss in detail about the requirements aspects of these systems and also how these systems can be made more reliable and secure via blockchain technology.

2.3.1 Patient data management

Patient data is the record of highly sensitive personal information about the medical history of the patient. It is inclusive of all the current and past medical issues, diagnosis, reports, treatment, and other medical procedures that a patient has gone through in his/her life cycle. It is essential to manage all the medical-related documents of a patient for upcoming future health-related issues and problems. With the help of the patient data, it becomes easier to understand the case better by knowing the patient's sensitivity and allergic equations and getting proper information of all past and ongoing treatments and problems. It becomes an essential necessity in the case of people with serious issues like chronic diseases, heart diseases, dialysis, and liver-related problems. Still, in countries like India, the people used to manage a paper-based file for all this related information, which is difficult to carry along with a person always. In emergency situations like sudden heart attack, it requires the patient medical data to be available to the doctors for the proper and right treatment. Due to the lack of availability of such information, various issues may arise like side effects of medication and improper treatments.

EHRs play an important role in such situations by making the data available to healthcare organizations and related persons, pharmaceuticals, insurance companies, and family members. The routine use of EHRs by clinicians produces rich patient-level data that can be applied to secondary uses like population health management, epidemic surveillance, and clinical, translational, and health services research. The value proposition of widespread adoption and meaningful use of EHRs, as promoted by numerous experts and professional organizations, lies not only in improving the quality of care and controlling costs but also in creating a healthcare system with "rapid learning" that can advance our knowledge in a variety of clinical and policy domains [19]. EHRs make it possible for medical researchers to collaborate, share their findings, and secure permission for data gathering and access. The lack of essential medical equipment and supplies during the COVID-19 epidemic exposed how susceptible healthcare workers are to supply chain disruptions. The administration of patient data is an intriguing use case for blockchain in the healthcare industry. Care coordination, data security, and interoperability issues are just a few of the issues that blockchain in healthcare can address by enhancing the accessibility, accuracy, security, and affordability of developing and maintaining electronic health records (EHRs) [20].

Blockchain innovations ensure information security by having command over delicate information and work with medical care executives and various experts of the clinical space for the patient. In the medical services settings, it can characterize an exchange as a course of making, transferring or moving patient's information that is performed inside the associated peers. A bunch of exchanges gathered at a specific time is added to the record that records all the exchanges and consequently addresses the condition of the organization. Medical Records (MedRec) [21] is a decentralized record management system that uses blockchain technology to handle the medical records of patients. As shown in Figure 2.6, the patients can without much of a stretch access their clinical data over clinical suppliers and treatment locales. MedRec oversees verification, secrecy, responsibility, and information sharing-critical contemplations while dealing with sensitive data. The key benefits of patient data management include:

- ***Scheduling appointment:*** From recent studies, it has been found that around 48% of people are booking their doctor's appointments via phone while 43% likes to book online. It makes it simple to book an appointment and check the availability of the doctor.
- ***Record of medical history:*** Patient data management avails to keep track, access, and store patient's medical data, such as demographic information, vaccination status, radiology images, etc.

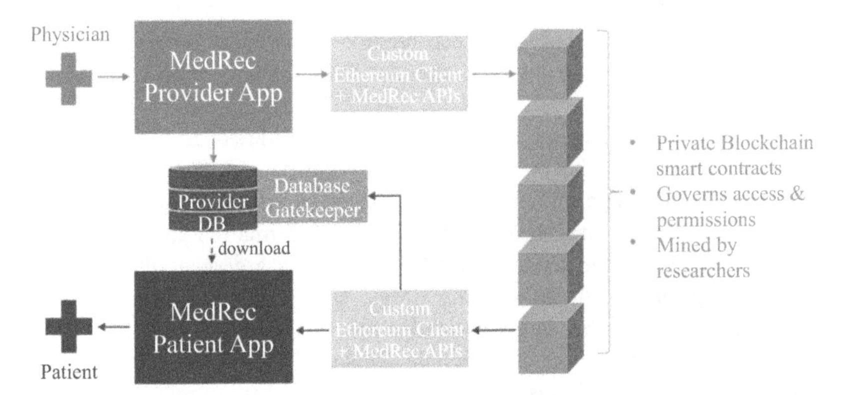

Figure 2.6 Prescription management system – work flow diagram

- ***Tracking all patient interactions:*** It keeps complete track of patient's visits to the doctor's clinic or hospital from check-in time to waiting time, patient's diagnosis, health concern, etc.
- ***Medical transactions and billing system:*** It monitors generally past and existing payment details. Alongside this, it records prescription subtleties for patients. The medical clinic may check the insurance eligibility of the patient, when a patient checks in.
- ***Predictive reports:*** The patient's progress reports help to predict the needs of the patient, such as his actions and lifestyle, hereditary conditions, current health status, etc.

2.3.2 Prescription management

Another important aspect of EHR management is prescription management. When it is talked about healthcare-related data, prescriptions given by doctors for particular health-related issues are considered as one of the necessary data. Now, the question arises why its need a prescription management system as one of the necessary e-healthcare services. This section explains the need of a prescription management system via reference to two cases (below mentioned) out of many more.

***Case* 1**: Sometimes the medicine prescribed by doctors is either running out of stock or not available at nearby medical stores of the patient. In such situations, either the patient or his attendant is required to revolve here and there to get that medicine, this whole procedure takes a lot of time and it might also cause the situation of the patient (having serious health issues) to get worse due to delay in the medicines. In order to avoid such a situation, it needs a strong prescription management and tracking system to get information about that medicine's availability and the related pharmacy's information [16].

***Case* 2**: Another case is when the patient is on some long-term or life-time medication, like sugar and BP patients. In that case, patients are required to keep a monthly stock of their medicines. Sometimes, due to their busy schedule, people use to forgot ordering their routine medicines. For such a situation, the prescription management system is required to keep records of previous orders and also it is required to notify/give a reminder to the customer for upcoming their upcoming order date.

A prescription management system can be considered as a web or mobile application that connect hospitals, pharmacies, and patient on a single platform. The health specialists can transfer a patient's solution information through the App while drug stores and e-drug stores can get to that information and send notices to said patient assuming the recommended prescriptions are accessible in the drug store. For long-term users of specific drugs, say those recommended to heart medicine, can be reminded to get their tops off. For patients, Android/iOS applications can be created and made accessible. This way, the patients can look at drug stores close to their area alongside medication accessibility. E-payment wallets can also be added to make secure payment transactions.

As shown in Figure 2.7, upon a medical clinic/hospital visit, the concerned specialist will transfer the patient's prescription information post-diagnosis. Using some reliable exchange method, the hash of this information will be added to the blockchain. Remembering that patient information is profoundly classified, it will be kept hidden and access will just be on a solicitation premise and empowered exclusively for concerned members. This will be accomplished through executing an answer, for example, a zero-information verification [17]. Health specialists will be able to see past clinical solutions and prescriptions of the patient with their authorization and continue appropriately. This will be particularly helpful when experts leading tests, analysis, or directing treatment need to know points of interest from the patient history like sensitivities or existing circumstances. Drug stores and E-Pharmacies will actually want to involve this information related to their stock reports and produce new reports in light of market interest. Information gathered through this framework can likewise be used for additional innovative work, while never compromising patient security.

The prime benefits of prescription management using blockchain include:

1. Wiping out the need of upholding prescription
2. Decreasing the number of orders getting dismissed because of indistinct or invalid prescriptions.

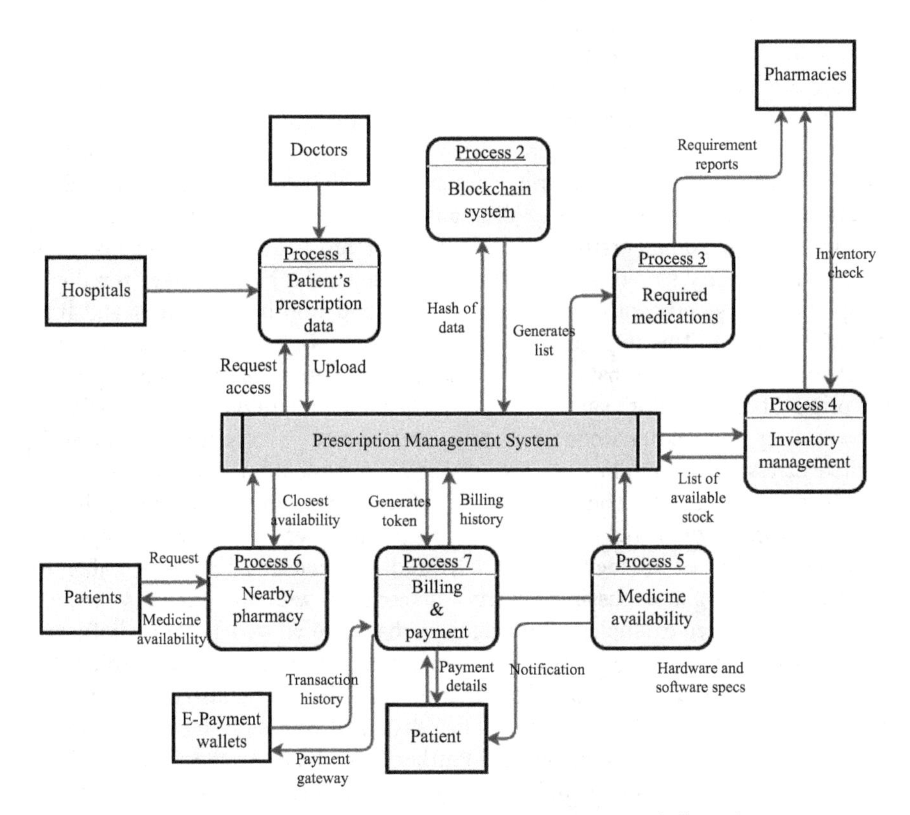

Figure 2.7 Prescription management system – work flow diagram

3. Empowers client to get all solutions from one stage
4. Kill substance misuse.
5. Valid and audited clinical records
6. Keep away from medication fakes

Subsequently, as obvious by the blockchain-based clinical prescription global positioning framework depicted above, better administration of information can tremendously further develop the medical services industry. In the exceptionally digitized world so dependent on an information-driven economy, blockchain can assist with making lives simpler.

2.4 Drug traceability using blockchain technology

Falsified medications are a major contributor to a major public health concern that has a severe effect on the lives of humans and the treatment response. The World Health Organization (WHO) characterizes fake drugs as items that are deliberately and deceitfully organization of meds with adoration to the provider and additionally personality. This definition applies to both prescription and over-the-counter medications. Falsified pharmaceutical products may have insufficient, incorrect, or erroneous ingredients, as well as falsified information, such as incorrect labelling and incorrect packaging. The WHO researchers estimate one out of ten medicines that are circulating in underdeveloped nations are either of questionable value or have been falsified. In advanced countries, approximately one to two percent of all the drugs that are devoured are counterfeit. The global exchange of fake medications host an impact on all gatherings associated with the drug business, including emergency clinics, drug stores, discount wholesalers, worldwide well-being developers, and administrative specialists. As a result of its participants using tainted, improperly stored, and falsified ingredients, the black market for illegal drugs makes a significant contribution to the production of counterfeit and fraudulent medicines. This is possible because there are not enough business or technical solutions available that provide an adequate solution for tracing and proving the origin of an item. For example, a bad-quality rendition of the counter malignant growth drug was sold and conveyed to a large number of disease patients in the United States, which might have brought about treatment-related confusion for a portion of these patients. The Asia-Pacific, African, and Latin American locales are the most helpless to the risks presented by counterfeit medications, which represent very nearly 30% of all medications delivered and consumed and are liable for practically 2.5 million passings yearly. In contrast with the earlier years, the quantity of revealed examples of medications that were viewed as phony in Europe has expanded by a variable of two. A prominent European research project has recently published a report that highlights the fact that the industry of fraudulent medicines is considered to be a more lucrative and profitable business than the process of selling legitimate medicines. Furthermore, the researchers estimated that a revenue loss equal to almost 6.5% in drug sales equates to €12 billion every year. It is becoming increasingly difficult to ensure the safety of products throughout the supply chain due to the increased availability of medications through unauthorized distribution networks and online pharmacies. Furthermore, restricted information permeability about stock and stock levels across the inventory network presents more noteworthy open doors for fakes to enter the market. This is because counterfeits are more difficult to detect. These measures must be put into place before the year 2023.

Blockchain offers a decentralized track-and-follow arrangement that is valuable for the observing of medications or clinical hardware from the source the whole way to the market racks. The decentralization highlight permits approved people to hold a confidential key to follow drugs over the inventory network. By and large, blockchain is utilized in a mix with different apparatuses, for example, IoT and PC vision for network and information move, and AI for independent direction. At each phase of the inventory network, blockchain makes it more straightforward to approve the legitimacy of your fabricated medications. Thus, blockchain-empowered start to finish observing permits you to forestall misfortunes coming from the tainting or robbery of bundles while they are being moved or bundled. With blockchain devices set up, your business can stay away from defilement-related drug gives that might harm your standing.

Blockchain innovation has presented another model of application advancement principally founded on the successful implementation of the information structure inside the Bitcoin application. The central idea of the blockchain information structure is like a connected rundown, for example, it is divided between all the nodes of the organization where every hub keeps its neighborhood copy of every one of the blocks (related to the longest chain) starting from its beginning block. Making a chain of blocks associated by cryptographic develops (hashes) makes it undeniably challenging to alter the records, as it would cost the modification from the beginning to the most recent exchange in blocks as shown in Figure 2.8.

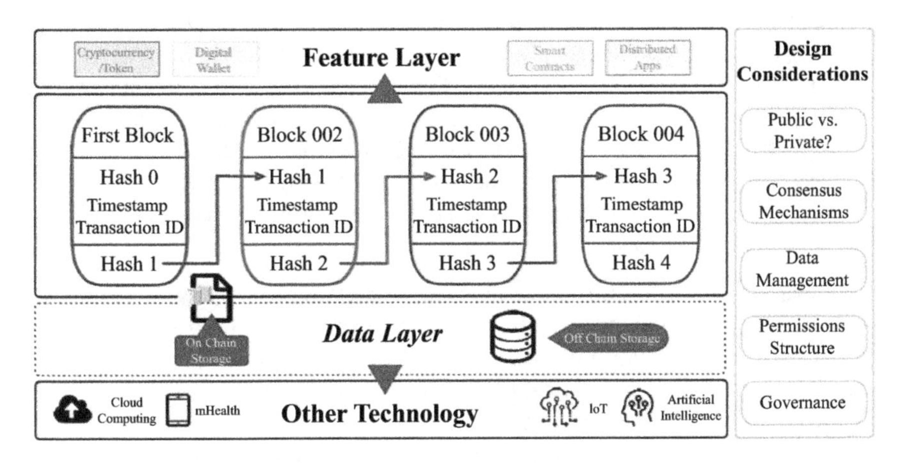

Figure 2.8 Blockchain information structure

2.4.1 Drug safety

Drug safety alludes to the recurrence of medication impacts (i.e., physical or research facility poisonousness that might actually be connected with the medication) that arise during treatment and were absent before treatment, or they become more regrettable during treatment contrasted and the pre-treatment state. Clearly, a medication (or any clinical treatment) ought to be utilized just when it will help a patient. Benefit considers both the medication's capacity to create the ideal outcome (viability) and the sort and probability of antagonistic impacts. Since the beginning of the previous hundred years, many demonstrations, regulations, or changes have been made to ensure that endorsed drugs are first protected and afterward powerful. Moreover, these guidelines are proceeding to change to ensure that these medications have a positive advantage risk balance. Customized medication ought to be thought about when drugs are given to patients in light of the fact that the pharmacokinetic cycle inside the body changes from one patient to another and starting with one explicit illness state then onto the next. Notwithstanding, unfriendly medication responses can be limited in the event that more safeguards are taken by medical care experts, particularly including the patient as one mainstay of the remedial arrangement and giving more quiet advising, which will further develop drug security [18].

2.4.2 Drug efficacy and effectiveness

The major factors on the basis of the drug safety of a medication that is measured are its efficacy and effectiveness.

- ***Drug efficacy:*** Efficacy is the ability to create a result like lower circulatory strain. The efficacy or viability of a medicine can be evaluated precisely just in ideal circumstances, i.e., when patients are chosen by appropriate rules and rigorously comply with the dosing plan. Hence, viability is estimated under master watch in a gathering of patients probably going to have a reaction to a medication, like in a controlled clinical preliminary.
- ***Drug effectiveness:*** Frequently, a medication that is solid in clinical preliminaries isn't extremely compelling in real use. For instance, a medication might have high viability in bringing down pulse, however may have low adequacy since it causes such countless unfriendly impacts that patients quit taking it. Likewise, the clinicians coincidentally endorse the medication improperly (e.g. giving a fibrinolytic medication to a patient ideal to have an ischemic stroke). Accordingly, viability will in general be lower than adequacy.

In the pharmaceutical industry, one of the most pressing concerns is thwarting the distribution of fake medications. The most likely explanation for this is that there is not currently an available protected digital platform that can be used to trace the origin of treatment pharmaceutical companies have really no way of understanding what is going on with their products even though data can be provided in between various systems and processes today. As a direct consequence, drug producers and pharma companies lose money because their products cannot be sold to customers.

Consumers also suffer the consequences of the situation. However, blockchain technology is the answer. You cannot alter the data, and there will always be a record of transactions that have taken place transactions can always be verified as legitimate because the software runs on different networks and is distributed. As a result of the availability of data that reveals the origin of the drug throughout the entire process of drug distribution, it will be exceedingly difficult for counterfeit medicine to make its way onto the market. This would also result in reduced costs associated with bringing pharmaceuticals to market [22].

2.4.3 Drug logistics

Logistics is one of the vital components of the ongoing business. It guarantees that things arrive at their expected spot in negligible time, cost, and harm. So, it manages the execution and association of the perplexing tasks for shipping products starting with one spot, and then, onto the next (for the most part from the starting place to utilization).

A pharmaceutical supply chain is organized in such a way that it follows a process from beginning to end, beginning with the sourcing of the active medication ingredients (source) and ending with the distribution and delivery of the final product (medication) to patients (end-users). Since the distribution of genuine and first-class products at the appropriate while has a direct impact on the patients' health and safety, it is the primary accountability of the members of the supply chain to fulfill this responsibility. The current distribution and administration systems for drugs have become significantly more extensive and complicated in recent years. Verifying transactions can be challenging due to a number of factors, including limited data visibility, unclear ownership structures, and a diverse group of stakeholders. Because there is not always a unified view of the production chain, it is frequently necessary to rely on centralized solutions provided by a third party in order to collect and validate the information.

2.4.4 Global challenges to logistics

Progressively globalized and complex supply chains are significantly affecting worldwide organizations. Partners with store network necessities to deal with an expanded measure of data while monitoring more exchanges, recording execution, and arranging future exercises.

Logistics processes including different parties require joint execution across each interaction step. Today most cooperation is led physically and disconnected, which frequently prompts redundancies and mix-ups. In this changing setting strategic capabilities face difficulties connected with sharing data along the store network safely in the following perspectives:

1. ***Transparency:*** The progression of data to help store network arranging and controlling (e.g., creation or dissemination).
2. ***Speed and efficiency:*** Getting the right merchandise to the perfect convergence of everything working out through digitized and productive processes (e.g., duty processes, approving the beginning of a transfer).

3. ***Traceability:*** Reconstructing the beginning and development of merchandise, and monitoring materials at each stage in the worth chain (counting verification of area, review trails furthermore, confirmations).
4. ***Payment:*** Transferring cash to providers proficiently and with dependable documentation.

2.4.5 Blockchain in managing drug logistics

In order to fulfill the continuous increasing demand, many pharmaceutical companies entered into mergers and acquisitions in order to scale and expand their global presence and markets. A drug may require to travel thousands of kilometers during its supply chain and during this cycle, these drugs are monitored by logistic companies to ensure their optimal handling, transport, and storage conditions. Neglecting to keep up with the right circumstances can adversely influence the viability of the items. Right now, the particular circumstances expected for medication are checked through shrewd gadgets like Internet of Things (IoT) all through the whole inventory network. As different members of the production network, for example, makers, strategies organizations, stores, and drug stores utilize their own records in IoT.

Blockchain can bring all partners of the inventory network onto a similar record and work with the following by presenting consistency and administration all through the inventory network. The intrinsic straightforwardness, changelessness and de-unified elements of blockchain guarantee adherence to all rules by any taking part authority. Any deviations from the consistence conditions can be overseen by making shrewd agreements that will caution important partners in the store network.

Figure 2.9 shows how a drug logistics system can be made more secure, authentic, and reliable using blockchain technology. Blockchain can be utilized to give clearness on exchanges and each and every other information that is being placed in blocks; with regards to a production network, blockchain innovation can be utilized to follow the progression of medications and production network tasks across borders. This intends that, at each step of the medication circulation process, blockchain applications can vouch for the realness of a shipment, subsequently making it almost incomprehensible for any sort of deceitful movement.

Following are the benefits of using blockchain technology in drug logistics:

- ***Quicker processes:*** Regular strategy processes include unending manual approving and documentation turnover. This is a tremendous issue in coordinated factors, and blockchain addresses it well. The Distributed Ledger Technology assumes control over all the check and endorsement-related problems. It additionally annihilates mistakes, and forestalls prohibited exchanges or unsanctioned organization changes. This rates up the work process since coordinated operations organizations can focus on other significant errands.
- ***Transparency:*** Transparency is key in keeping up with trust among circulation accomplices. Sadly, it is likewise a major worry in planned operations. By benefiting blockchain improvement administrations, operations organizations

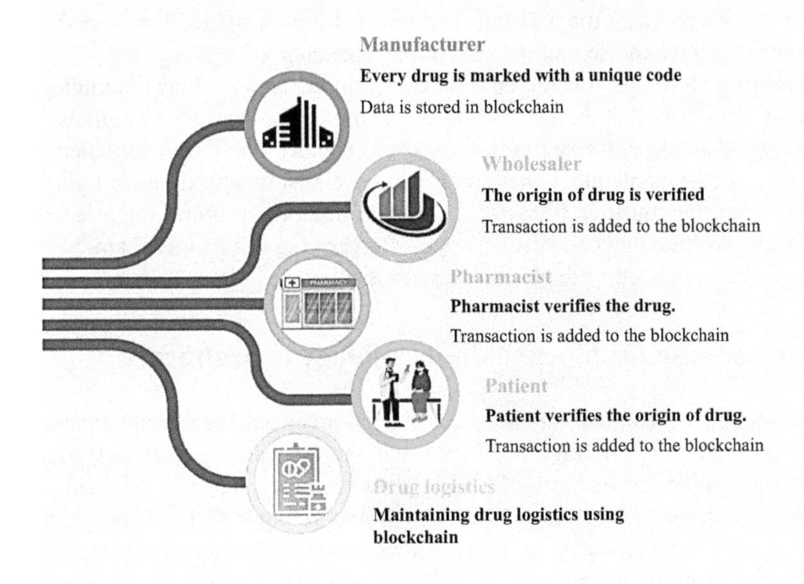

Figure 2.9 Drug logistics: supply chain management

can guarantee admittance to each snippet of data in the organization through blockchain. Subsequently, every exchange stays straightforward and changeless. This mitigates plausible trust-related pressures among organizations, something conventional strategy tasks neglect to achieve.

- ***Decentralization:*** The traditional methods engaged with coordinated factors permit organizations to confine admittance to their information as and when they see fit. This frequently makes questions among parties because of moral worries around information falsification. Decentralization diminishes the possibilities of misrepresentation by taking out the job of any incorporated power. This is conceivable because the same length of information allotment stays equivalent and steady among record clients. Along these lines, information restraining infrastructure becomes unthinkable, further developing planned operations.
- ***High efficiency via smart contracts:*** Smart contracts are programs present on the blockchain that actuate when the information meets specific pre-chosen conditions. One case of a savvy agreement might include the arrival of an instalment to the provider, relying on the prerequisite that the shipment shows up at the assembling premises.
- ***Data security:*** A blockchain offers momentous security since one block interfaces with different blocks from the two closures. Outsiders cannot adjust that information because they do not have the power. So, a focal authority must

be present, that works on the security for coordinated factors. Clients can screen the source with the assistance of information assembled that reveal an insight into merchandise and the executive producing processes.

- ***Scalability:*** All things considered, blockchain innovation is almost boundless. You can anticipate that there should be zero clogs, regardless of the number of exchanges that add to the strategic organization, which implies no bottlenecks! Choosing a reasonable blockchain improves specialist organizations in light of the fact that prerequisites fluctuate contingent upon the venture. Experienced organizations know this and are equipped for carrying out a blockchain-based arrangement, adjusted to the business methodology.

2.5 The use case for blockchain technology in healthcare

Although blockchain is a relatively new technology, it can help healthcare enhance its current methods for handling health data and supply chain transparency when combined with other technologies like big data analytics, Internet of Medical Things (IoMT), and 5G. Healthcare firms are collaborating with blockchain consultants to speed up the adoption of this technology in healthcare projects as blockchain gains prominence. Seven of the prominent blockchain technology applications in healthcare are discussed in subsequent subsections.

2.5.1 Clinical trials

Every transaction can be verified because the blockchain logs every change that occurs across the entire peer-to-peer network. This is one of the technology's most advantageous characteristics. Because clinical trials call for an efficient and open method of locating patient data, blockchain technology may eventually become the answer to this problem. The most open and honest clinical practices are the most effective. This includes both the individual who enters the information and the location where the information is entered. Audit trails are the only method currently available for viewing this data, but blockchain technology has the potential to enhance both the privacy and security of these processes. On the other hand, as is typical in the healthcare industry, there are a few questions regarding the applicable regulations that have arisen. Due to the extremely sensitive nature of clinical trial data, its protection is essential. Depending on the preferences of the user or the company, blockchain has the capability to either make all data completely visible or to make all data completely password protected.

In the field of healthcare, blockchain technology has the potential to act as a substitute for existing distributed database management systems. These systems have traditionally taken the form of client–server databases that accept Structured Query Language or relational input. Traditional distributed database management systems are an established platform in health-care systems; however, they have substantial limitations, such as the inability to support peer-to-peer data sharing, the susceptibility to external adversaries (e.g., hacking), and the absence of an immutable (that is unchangeable) audit trail. Blockchain may be able to solve these

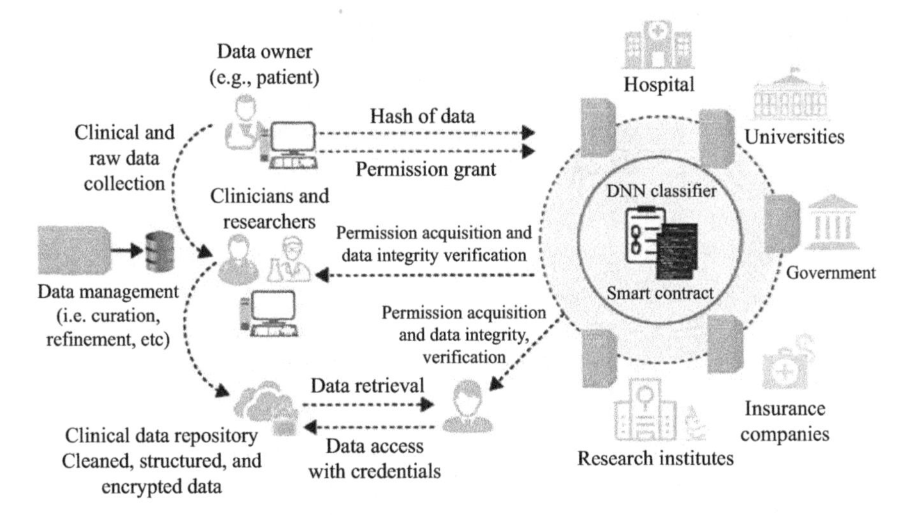

Figure 2.10 Application of blockchain in digital healthcare system

problems due to the unique properties it possesses, such as transparency, traceability, non-repudiation (the inability to dispute the validity of a signature), disintermediation (the removal of intermediaries from a decision-making process), and immutability. This would enable it to provide significant advantages over traditional platforms (e.g., distributed database management systems). A decentralized healthcare data management system could be created with the help of blockchain technology. This system would be responsible for coordinating on-chain events, which are transactions that are recorded on the blockchain ledger, and off-chain events, which are events that take place outside of the blockchain and are typically too large to store on the blockchain. Figure 2.10 is a representation of the construction of a blockchain-based healthcare ecosystem amongst a variety of stakeholders.

2.5.2 Smart contracts

Smart contracts can produce remarkable effects in the health insurance industry in particular. The suggested smart contract framework for electronic health records is summarized in Figure 2.11. The parameters of the agreement are defined in this smart contract, as well as how users interact with the Ethereum blockchain. The registration of healthcare institutions on the application is the responsibility of the administrator. Hospital accounts can be created, modified, added to, or suspended by the administrator. A new block is formed at the back end when a new patient record is created, confirmed, and broadcast to all the network nodes. The blockchain now has a new block. After a patient has been sent an access request and consented to access, the system is set up to allow hospitals to read and produce patient medical data. When a patient has to be seen by a different doctor, the hospital makes the request. The hospital's employees cannot access the patients'

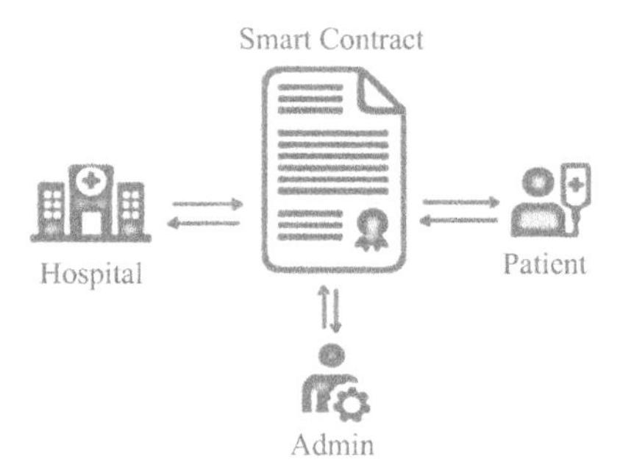

Figure 2.11 Basic smart contract framework

medical records if a request is denied. Patients can also make an appointment with the hospital. Two actions by the patient, i.e., permitting or rejecting, access to medical records and seeing appointment schedules – define the conditions of contact.

As blockchain is utilized as an enabler technology to share records, this architecture assures that data transmission is simpler and that data confidentiality, privacy, and integrity are preserved. The conventional insurance system may be greatly improved, and all pointless middlemen can be removed. When a patient uses smart contracts to purchase their medical insurance, the policy information is stored in the patient's blockchain profile and is less vulnerable to hacking than when it is kept in a conventional database. Platforms using smart contracts can cut expenses by removing middlemen from the payment chain. When specific criteria are satisfied, a smart contract (a series of if-then-else expressions) that is put on a blockchain is automatically executed. Smart contracts start the workflow for completing transactions without the need for an intermediary. On a digital ledger, smart contracts make it easier to save medical records. With this configuration, the patient can switch doctors and hospitals without having to transfer their medical information and fill out a number of forms to retrieve their data, which exposes them to errors and jeopardizes the security of their records. Only the permissions for access change with smart contracts. The patient should ideally be able to enter her private key at any clinic of her choosing (including one that is in another nation), granting that facilitate access to his/her medical records. This is an example of blockchain interoperability in action, and while it is still in the early stages, many businesses are working to advance this idea.

In addition to all these use cases, among the most considerable opportunity afforded by blockchain technology is the increase in precision medicine. To become more specific, patient-centric identity grants the patient rights to agree to the use of their own data in interaction for health services or even recompense, and the choice about how to use their information. This is in addition to the patient having control over how their data is used. All in all, the genuine worth of a blockchain for the drug business probably would not be found in specific use cases.

2.6 Challenges to implement blockchain technology in e-healthcare record management

A blockchain network is a distributed ledger that can be accessed by all stakeholders in the pharmaceutical supply chain. Store their most important business data, and all users of the platform have access to the sensitive and private information that is stored there. It is possible that potential stakeholders will be hesitant to participate in such networks due to the fact that doing so could result in the loss of their competitive advantage [23]. This is especially true in situations where multiple businesses compete for customers in the same market. Since blockchain is a brand-new technology, its full potential has not yet been realized. It faces a number of obstacles in the healthcare industry, some of them are listed below.

1. ***Lack of technical knowledge***: Not everyone using the service will have access to sophisticated computers and programs. Furthermore, a large portion of consumers simply is not up to speed on today's most cutting-edge tools. Numerous senior citizens, for instance, are not comfortable with modern technology and so do not own or regularly use laptops or PCs. To mine cryptocurrencies, a graphics processing unit (GPU) is needed, and this is not standard in laptops. That's a major difficulty with blockchain technology.

   Solution: The end users need to be updated with technological developments. The media should help disseminate these cutting-edge tools. Businesses should also work to lower the costs associated with these innovations.
2. ***Lack of paperless method adoption***: While electronic health records have become commonplace, many consumers and clinicians still prefer paper records. They favor using a filing system to store patient information. It is not all paperless pharmacies yet. Prescriptions are the standard method used by pharmacies for tracking stock. It is helpful for patients to maintain the documents on hand, too. Consequently, transitioning to a blockchain-based, paperless system is difficult.

   Solution: This problem can be solved by having doctors practice on miniature blockchain. They need to use as few papers as possible. Paper can be used for X-ray plates and other surgical records. However, blockchains should be adopted by the healthcare sector for the storage of prescriptions and other file records.
3. ***Lack of government involvement***: In the United States, the government owns the vast majority of hospitals. This means the government must become involved in order to enforce regulations. Certain governments are keen on embracing cutting-edge innovations. Since Blockchain is a distributed ledger that relies heavily on its network of nodes, it would be impossible to implement it in publicly funded healthcare facilities. There is no governing body or outside actors to weigh in on matters. Blockchain is often used to make choices.

   Solution: The government should implement blockchain technology in healthcare settings as well. Blockchain's distinctive mix of safety and decentralization makes it stand out. Data preservation enables the government to store information for later use in studies and other endeavors.

4. ***Lack of cost reduction***: The use of blockchain technology still carries a high price tag. Complex and pricey software and gear are needed for mining. Mining cryptocurrency requires high-priced graphics processing units. Each block must also have its own dedicated electrical supply in order to be mined. Miners are not adequately compensated, though. This is why the price tag for this technology is so high.

   Solution: Blockchain businesses should prioritize lowering operating expenses as technology improves daily. Companies could also provide additional incentives for miners to participate in the blockchain.
5. ***Lack of privacy***: Blockchain's database stores all of this data and each user has a copy of it, so even if a node in the network goes down, the information is still secure and can be updated at a later time. Many customers would rather not disclose their health issues publicly. Therefore, it interferes with people's right to privacy.

   Solution: Comparatively modest private blockchains can be employed for this. A private blockchain can only be used with the approval of the network's operator. In this way, only anxious patients need to apply to join the network.
6. ***Lack of incentive***: Incentives are given to miners if and only if they verify a block. Blockchain technology allows for greater autonomy with one's finances, yet the incentives are not sufficient. This motivates them to mine more in the hope of increasing their total. When using a powerful machine, you can mine one bitcoin in as little as 10 minutes. Given the cost of the necessary hardware and software, the incentives generated are negligible. Blockchain's incentives may be lower than those of traditional systems, but businesses are finding ways to increase its value all the same. Companies have established a wide variety of incentive programs in recent years.
7. ***Lack of cryptocurrency acceptance***: A large majority of medical professionals do not accept bitcoin payments. Online payment was not well-established prior to the introduction of Blockchain technology. The money is still coming in. Therefore, it is necessary to create a reliable system for accepting payments online. Blockchain technology then has to be implemented.

   Solution: This problem can be solved if the government recognizes cryptocurrency as a legitimate form of payment. Similarly, healthcare facilities should accept cryptocurrency payments.
8. ***Lack of cyber security***: Despite Blockchain's robust security and the absence of trusted third parties, threats like the 51% attack, the Sybil attack, and others have emerged as a serious concern. Many hackers aim for users' wallets because they contain a lot of cash. Hackers often try to boost their earnings by generating network congestion and mining blocks that aren't needed.

   Solution: The hospitals require to authenticate user using unique IDs, eliminating any potential security risks. Utilizing private Blockchains allow the network administrator to validate each user.
9. ***Lack of central healthcare***: In most countries, medical care is provided through decentralized networks. Multiple locations are home to many hospitals. So, it is clear that keeping up with a blockchain is a demanding job.

Adopting blockchain technology for medical records management would be impossible without a centralized database. Consider the case of someone who goes to a hospital labelled "A" in state x. He may return to Hospital "A," but this time to a different branch. Each geographical region has its own blockchain. This means that getting your hands on your old medical records will be a major hassle.

Solution: The hospitals may implement blockchain technology in certain areas so that problems may be identified and corrected. Distributed networks are preferable, and it is recommended that each hospital have its own huge blockchain for information storage.

10. ***Lack of speed***: Blockchain technology is quite slow. If the network is particularly vast, the processing time can be prohibitive. Due to the lengthy confirmation processes, communication speeds are reduced.

    Solution: The standards for the maximum block size of a blockchain should be set appropriately. Blockchains should have an appropriate size, falling between the extremes. It is important to maintain a system to prevent the placement of unwanted roadblocks. A rigorous testing regimen and the elimination of needless traffic are also responsibilities of the technical staff.

Experimentation with blockchain technology has been going on for the past two years, with varied outcomes so far due to the facts of need to collaboratively start sharing and preserving the confidentiality of healthcare data. Despite this need, the experiments have produced mixed results so far. This phase of exploration of the possible applications of blockchain technology in the healthcare industry has been characterized by the rise and fall of concrete evidence and pilot developments that have not entered into robust manufacturing and are still not being used frequently. On the other hand, there is now an increasing "consensus" and progress being made on use cases that are considered viable for blockchain and health. Many use cases include patient-centered identity and drug supply chain management, as well as clinical trials and research [24]. In the medical care industry, information storehouses that are supplier driven as opposed to patient-driven can be settled by empowering open well-being information trade advertisements that are driven by patients. Clinically checked diagnostics, treatment results, genuine proof, hereditary qualities, DNA profiling and more will make up the main part of the market information. Many previously unimaginable possibilities are now possible thanks to a growing open health data market.

However, technical data requirements and regulatory measures must be developed in order to promote compliant as well as effective blockchain implementation at massive scales in life health and science applications. These really are essential as they will ensure that the allocation, management, and control of patient data are governed by appropriate protocols and policies. Government controllers are significantly more prone to embrace arrangements and suggestions for advancements with market-driven and agreement-based specialized guidelines. Despite the fact that administrative offices don't foster specialized principles and administrative organizations do not foster approach, administrative organizations are significantly more liable to embrace strategy and direction for advancements with market-driven and agreement-based specialized guidelines. Here, most critical association lies between policymaking and normalization.

In reality, new technologies will not be adopted for manufacturing or commercial devices if they are launched in the market prior to establishing credibility. Thus, business and consensus-based technical specifications specifically address some of the apprehensions surrounding the adoption of new technologies, such as blockchain. These include sponsoring interoperability with the already existing software systems, reducing assimilation costs by eliminating the need for customization, establishing credence (through consensus-built standards), enabling industry-wide implementation, and continuing to drive competition through the availability of new stand-alone applications (where possible).

As anticipated, industry executives, scientists, and engineers are focusing on the present, particularly on the successful implementation of a fully functional blockchain for just a specific use case. Nevertheless, organizations that establish standards must place their primary emphasis on developing solutions for future problems. The introduction of multiple blockchains will necessitate the institution of supplemental system integration, the provision of insights to government regulators in order to formulate principles and regulations, and the maintenance of continuous education software for industry players and patients.

Blockchain technology has not yet been fully utilized to its full potential. However, the results of blockchain pilot projects, both those that are successful and those that are not, will eventually lead to the fulfillment of the promise of patient-driven healthcare systems in the form of open health data markets and precision medicine, which will then reach patients. Pharmaceutical companies are able to eliminate the problem of selling fake medicines, thanks to this system, which also contributes to a significant increase in business. As part of this ongoing work, the whole plan is to expand the size of the network and put it into action in real-time within pharmaceutical companies in order to evaluate its efficacy and ensure its validity. In addition to this, the work may also apply specific machine learning models' recommendation outputs to enhance the accuracy and effectiveness of the system [25].

2.7 Conclusions and future scope

This study proposes a distributed traceability solution for the management of healthcare records that uses IoT and Blockchain technologies to make sure that healthcare data can be tracked in a smart way. In this chapter, the author suggests a complete solution with a five-layer blockchain platform architecture that works on design, development, application, and evaluation for each layer. The first layer of the system is made up of electronic health records. The second layer is a system for managing patients. The third layer is for managing prescriptions. The fourth layer is for tracking drugs, and the last layer is for making sure drugs are safe. The proposed system uses blockchain technology to deal with issues like management, maintenance, security, usability, and tracking.

One of the main goals of our proposed plan is to try to set up a system that works with blockchain technology and machine learning. The system for managing the drug supply chain and the system for managing drug recommendations are the two parts that make up this system. The use of machine learning algorithms and blockchain technology in the medical field has led to great results. Throughput, the time it takes to respond to a money transfer, and the cost of communication are all examples of performance measures that have been used in the many tests that have been done to see how well our system works. Scholars from Asia and other parts of the world should pay more attention to blockchain-based theory in the future.

References

[1] S. Yu, J. Lee, K. Lee, K. Park, and Y. Park, "Secure authentication protocol for wireless sensor networks in vehicular communications," *Sensors*, vol. 18, no. 10, 3191, 2018.

[2] E. L. Eisenstein, R. Collins, B. S. Cracknell, *et al.*, "Sensible approaches for reducing clinical trial costs," *Clinical Trials*, vol. 5, no. 1, pp. 75–84, 2008.

[3] I. Yaqoob, K. Salah, R. Jayaraman, and Y. Al-Hammadi, "Blockchain for healthcare data management: opportunities, challenges, and future recommendations," *Neural Computing and Applications*, vol. 34, no. 1, pp. 1–16, 2021.

[4] H. Tao, M. Z. A. Bhuiyan, A. N. Abdalla, M. M. Hassan, J. M. Zain, and T. Hayajneh, "Secured data collection with hardware-based ciphers for IoT-based healthcare," *IEEE Internet of Things Journal*, vol. 6, no. 1, pp. 410–420, 2018.

[5] M. Aazam, Z. Sherali, and A. H. Khaled, "Health fog for smart healthcare," *IEEE Consumer Electronics Magazine*, vol. 9, no. 2, pp. 96–102, 2020.

[6] F. Ali, S. El-Sappagh, S. M. R. Islam, *et al.*, "A smart healthcare monitoring system for heart disease prediction based on ensemble deep learning and feature fusion," *Information Fusion*, vol. 63, no. 1, pp. 208–222, 2020.

[7] M. Bhatia, S. Bhatia, M. Hooda, S. Namasudra, and D. Taniar, "Analyzing and classifying MRI images using robust mathematical modeling," *Multimedia Tools and Applications*, vol. 81, no. 26, pp. 37519–37540, 2022.

[8] S. Namasudra and P. Sharma, "Achieving a decentralized and secure cab sharing system using blockchain technology," *IEEE Transactions on Intelligent Transportation Systems*, pp. 1–10, 2022. DOI:10.1109/TITS.2022.3186361.

[9] A. Singh, A. Kumar, and S. Namasudra, "DNACDS: Cloud IoE big data security and accessing scheme based on DNA cryptography," *Frontiers of Computer Science*, vol. 18, no. 1, 181801, 2024. DOI: https://doi.org/10.1007/s11704-022-2193-3.

[10] A. Azaria, A. Ekblaw, T. Vieira, and A. Lippman, "MedRec: Using blockchain for medical data access and permission management," In *Proceedings of the 2016 2nd international conference on open and big data (OBD)*, IEEE, Vienna, Austria, 2016, pp. 25–30.

[11] J. Zhang, N. Xue, and X. Huang, "A secure system for pervasive social network-based healthcare," *IEEE Access*, vol. 4, no. 1, pp. 9239–9250, 2016.

[12] T. T. Kuo, H. E. Kim, and L. Ohno-Machado, "Blockchain distributed ledger technologies for biomedical and health care applications," *Journal of the American Medical Informatics Association*, vol. 24, no. 6, pp. 1211–1220, 2017.
[13] S. Angraal, H. M. Krumholz, and W. L. Schulz, "Blockchain technology: applications in health care," *Circulation: Cardiovascular Quality and Outcomes*, vol. 10, no. 9, p. e003800, 2017.
[14] A. Dubovitskaya, Z. Xu, S. Ryu, M. Schumacher, and F. Wang, "Secure and trustable electronic medical records sharing using blockchain," In *Proceedings of the AMIA Annual Symposium Proceedings*, American Medical Informatics Association, 2017, p. 650.
[15] The HIPAA Privacy Rule: U.S. Department of Health and Human Services. Available: http://www.hhs.gov/hipaa/, 2017 [Accessed on 13 August 2022].
[16] I. Bajrovic, M. D. Le, M. M. Davis, and M. A. Croyle, "Evaluation of intermolecular interactions required for thermostability of a recombinant adenovirus within a film matrix," *Journal of Controlled Release*, vol. 341, no. 1, pp. 118–131, 2022.
[17] Code of Federal Regulations. Available: https://www.hhs.gov/ohrp/regulations-and-policy/regulations/45-cfr-46/, 2017 [Accessed on 13 August 2022].
[18] PROVISIONS, CHAPTER I. GENERAL. "Directive 95/46/EC of the European Parliament and of the Council on the protection of individuals with regard to the processing of personal data and on the free movement of such data." *Official Journal L*, vol. 281, no. 23/11, pp. 0031–0050, 1995.
[19] K. Zheng, Q. Mei, and D. A. Hanauer, "Collaborative search in electronic health records," *Journal of the American Medical Informatics Association*, vol. 18, no. 3, pp. 282–291, 2011.
[20] F. K. Nishi, M. M. Khan, A. Alsufyani, S. Bourouis, P. Gupta, and D. K. Saini, "Electronic healthcare data record security using blockchain and smart contract," *Journal of Sensors*, vol. 2022, no. 1, pp. 1–22, 2022.
[21] P. Sharma, M. D. Borah, and S. Namasudra, "Improving security of medical big data by using Blockchain technology," *Computers & Electrical Engineering*, vol. 96, no. 1, p. 107529, 2021. DOI: https://doi.org/10.1016/j.compeleceng.2021.107529.
[22] S. Rajpoot, M. Alagumuthu, and M. S. Baig, "Dual targeting of 3CLpro and PLpro of SARS-CoV-2: a novel structure-based design approach to treat COVID-19," *Current Research in Structural Biology*, vol. 3, no. 1, pp. 9–18, 2021.
[23] Z. Wang, L. Wang, Q. Chen, L. Lu, and J. Hong, "A traditional Chinese medicine traceability system based on lightweight blockchain," *Journal of Medical Internet Research*, vol. 23, no. 6, p. e25946, 2021.
[24] Y. Lu, Y. Shi, and J. You, "Strategy and clinical application of up-regulating cross presentation by DCs in anti-tumor therapy," *Journal of Controlled Release*, vol. 341, no. 1, pp. 184–205, 2022.
[25] M. Hasannia, A. Aliabadi, K. Abnous, S. M. Taghdisi, M. Ramezani, and M. Alibolandi, "Synthesis of block copolymers used in polymersome fabrication: application in drug delivery," *Journal of Controlled Release*, vol. 341, no. 1, pp. 95–117, 2022.

Chapter 3

Blockchain in pharmaceutical sector

Pratima Sharma[1], Madhuri Gupta[1] and Gagandeep Kaur[1]

Abstract

The pharmaceutical companies that manufacture, ship, and supply medical products face various challenges in tracing the products and identifying the counterfeit drugs in the supply chain system. Counterfeit drugs are one of the major consequences within the existing supply chain, which seriously affects human health and causes severe loss to the economy of the healthcare industry. Moreover, the traditional supply chain systems suffer from various challenges such as lack of transparency, limited tracing and tracking of products, privacy-related issues, and inaccurate information. The solution to the identified challenges is to leverage blockchain technology in the pharmaceutical system that adds traceability, visibility, privacy, and security to the drug supply chain. Blockchain technology is an advanced technology that stores medical product-related details in a completely decentralized manner and ensure transparency in the supply chain process. Thus, this chapter presents the drug discovery process, the pharmaceutical supply chain management system and its challenges, the need for clinical trials, and various preventive measures involved in the current pharmaceutical supply chain consortium. Then, we discuss the use case of a blockchain-based supply chain system for the pharmaceutical sector to track medical products and highlight its functionalities using the smart contract concept. The use case also explains the various smart contract concepts utilize to guarantee data provenance, eliminate the intermediatory authority, provide an immutability feature with security, maintain product traceability, and ensure transparency.

Keywords: Blockchain; Supply chain; Pharmaceutical; Drug counterfeiting; Healthcare

3.1 Introduction

The healthcare supply chain system is a complex network with multiple shareholders such as the researchers, manufacturers, distributors, pharmacists, hospitals,

[1]Department of Computer Science and Engineering Bennett University, India

and patients. Tracing and tracking the product or raw materials using the supply chain network is difficult due to several factors like inadequate information availability, centralized model, and competing entities. According to the world healthcare organization report, counterfeiting drugs has been considered the most common concern. Counterfeit drugs are knowingly and deliberately produced or mislabeled by the fraud stakeholders of the supply chain system to make them appear as genuine products [1,2]. It is estimated that every tenth drug in the market circulation is found to be counterfeited or of poor quality. Thus, the utilization of poor-quality drugs may impact healthcare products, negatively affecting the mortality rate. Moreover, tracking of drugs is an important feature in the supply chain system that ensure the safe and trusted supply of drugs in an acceptable condition. The drug life cycle begins from the manufacturing that produces batches of drugs. Next, the manufactured batches of drugs are processed by various entities to convert them into medical products and end the cycle with the traders that sell them to customers. During the entire drug supply chain cycle, the ownership of the drugs changes until the final product supply to the user with the complete record of the drug from its origin to the final delivery.

The developments in cryptography and distributed computing have introduced an advanced technology called a blockchain. Blockchain is a distributed ledger that replicates and exchanges data through peer-to-peer networks. Blockchain was initially introduced by an unknown person, Satoshi Nakamoto, who created Bitcoin to trade digital currencies directly without third parties [3]. Nakamoto developed the paradigm of a network of nodes working to maintain a decentralized and secure database. The blockchain is the core technology behind cryptocurrencies – a shared public database or a continuously updated registry of all transactions. Blockchain can be regarded as a technical breakthrough and financial advancement [4,5]. It provides a solution to any problem using a trustworthy ledger in a decentralized setting where it is impossible to completely trust actors, humans, and computers. In addition, the blockchain follows a series of procedures and cryptographic mechanisms applied to a shared network to secure data storage within a distributed database composed of authenticated blocks encapsulating the data. It stores the data as an ordered list of blocks. By referencing the previous block's hash, each block distinguishes by the hash sequence and tied to the preceding block. The only anomaly is the first block (called "Genesis block"), which does not have the previous block's hash value, known as the ancestor block. The data is passed to the miners, who verify it by solving mathematical puzzles and attaining consensus. The three key principles ensuring the system's functionality are (1) blocks and hashing, (2) mining, and (3) consensus [6].

Therefore, the advantages of blockchain technology provide transparency and trust in the pharmaceutical supply chain system. The blockchain-based supply chain in the pharmaceutical system increases the visibility of the global supply chain process with the control of counterfeiting. Various researchers proposed the many models that utilize blockchain technology in the supply chain system to increase the efficiency of the supply chain process in the pharmacy sector. For example, Musamih *et al.* [7] propose a blockchain-based drug traceability system in the healthcare domain using the Ethereum blockchain framework. The proposed system implements the supply chain functions to achieve traceability in the healthcare supply chain system without involving any trusted third party. It guarantees data security, privacy, transparency, and the elimination of counterfeiting. Subramanian *et al.* [8] present an integrated mobile application using blockchain and a pharma supply chain system with cryptocurrency. The presented application is implemented using the NEM blockchain network that utilizes XEM cryptocurrency for purchasing medical products. Similarly, Alkhoori *et al.* [9] designed a blockchain-based vaccine distribution system with a smart shipping process. It utilizes the Ethereum blockchain network for the implementation of cloud services. The designed architecture monitors the sensitive shipments using the smart container by recording the vital sign of shipments. The cloud services store monitoring-related details that provide the global access data feature. Therefore, blockchain technology ensures transparency, security, traceability, etc. and provides a robust supply chain system. This chapter further explores the different areas of the pharma supply chain system and, finally, presents the details of the blockchain-supply chain system in the pharmacy sector with the help of the use case.

The main contributions of the chapter are given below:

1. The chapter provides the complete details of the drug discovery process along with the explanation of various steps such as the early discovery process, preclinical phase, approval process, and review step.
2. It presents a detailed explanation of the pharmaceutical supply chain management system and identifies its challenges.
3. The chapter also discusses the supplier consortiums and preventive measures. It includes the details of revolutionizing clinical trials using blockchain technology.
4. It explains the concept of implementing blockchain technology in the pharmaceutical supply chain system with the help of product tracking use case.

The rest of the book chapter is organized as follows. Section 3.2 covers the complete details of the pharmaceutical drug discovery process. Then, the discussion of the pharmaceutical supply chain management process is presented in Section 3.3. Section 3.4 explains the concept of supplier consortiums and the various preventive measures. Next, the details of clinical trials and the need for traceability concepts are given in Section 3.5. Section 3.6 provides the details of a blockchain-based pharmaceutical supply chain system with the help of a use case. Finally, the chapter concludes in Section 3.7.

3.2 Pharmaceutical drug discovery

Pharmaceutical drug discovery is the process of identifying new therapeutic entities. Drug discovery is a long and complex process that may take several years. Drug discovery is so much time-consuming because of the various stages that an active ingredient has to go through and finally make it to the market. Also, it takes thorough examination and intensive research to authenticate an active compound as a safe drug. The process of drug discovery process starts by screening active

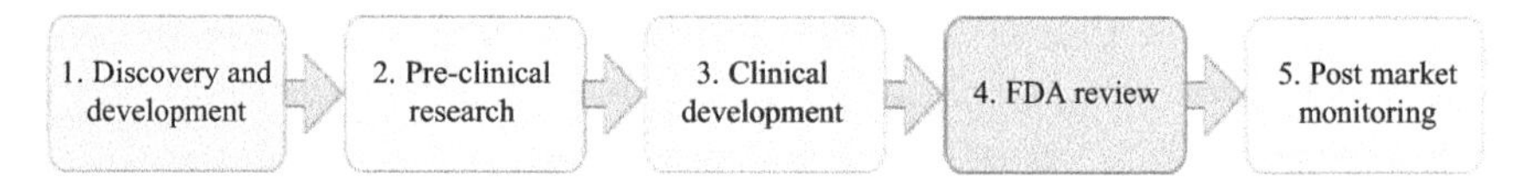

Figure 3.1 Drug discovery process

compounds as shown in Figure 3.1. These active compounds follow a lifecycle from discovery to approval. This active compound is tested and examined to be authorized as a safe drug ready to be marketed. Thus, the drug discovery process helps in the introduction of novel therapeutics and vaccines into the market. The whole journey of drug discovery is a costly process. A Tufts Center study states that it costs around $2.6 billion for the whole lifecycle of drug discovery. Furthermore, a drug's development and maintenance costs are around $312 million. Thus, it takes around three billion per drug in the entire research and development process of drug development [10, 11]. In recent times, further to the aspect that drug discovery requires large investments, it has turned to be more cumbersome. Typically, only 1 out of 10,000 drugs reach the market after approvals. Also, only 250 out of 10,000 candidate drugs pass the regressive pre-clinical testing. Thus, pharmaceutical drug discovery can be divided into five processes discovery and development, pre-clinical research, clinical development, FDA review, and post-market monitoring.

3.2.1 Discovery and development

The discovery and development process involves the process of discovery of new medications. In this process, research for a new drug begins in the laboratory. The aim is to produce the desired effect on a specific disease for its treatment. The initial stage involves the discovery of active ingredients. It involves the investigation of chemical libraries that include small molecules or plant extracts and discovering those have therapeutic effects. In this process, various mechanisms are involved, such as screening hits, iterative medicinal chemistry, and optimization of hits. These mechanism helps to diminish the potential drug side effects. This process involves different procedures and testing. Researchers collaborate to conduct an experiment using biochemical assays, cell cultures, silico platforms, and various animal models [12]. Target identification is the primitive step in this process wherein the protein or gene identified plays a prime role in disease. High-throughput screening (HTS) is an automated process in which searches for "hits" is conducted. The screening evaluates a particular target, such as a protein. It involves target identification, where the therapeutic target or molecular biological structures are identified that are "Druggable." A therapeutic molecule should "hit" and modify the biological activity. The target identification is obtained by tracking available databases. A molecule that interacts with the previous target to achieve the desired therapeutic effect is a hit. High content screening, phenotypic screening, fragment-based screening, structure-based screening, and virtual screening are strategies used to discover hits. The screening process results in a product termed as "lead." The Hit-To-Lead (H2L) process helps to find the pathway to a clinically active drug. Researchers refine the initial compounds using different screening methods, such as high-throughput screening, affinity selection of large chemical libraries, fragment-based techniques, and target-focused libraries. After this, lead optimization is performed to improve effectiveness, lower toxicity, or increase absorption [13]. A druggable entity needs to be validated to prove its therapeutic effect. The various technologies involved in target identification and validation are data mining using bioinformatics, genetic association, phenotypic analysis, and pathway approach. In recent times, target validation primarily involves disease association, bioactive molecules, cell-based models, protein interactions, signaling pathways analysis, functional analysis of genes, in vitro genetic manipulation, antibodies, chemical genomics, etc., [14]. The Assay development plays a major role in the discovery and development process. An assay is a test system that analyses the effect of the drug candidate on various levels such as cellular, molecular, and biochemical.

3.2.2 Pre-clinical research process

After the target identification of the potential drug, the researchers perform the next process, which is pre-clinical testing. This pre-clinical process begins once the lead compound is found. In the vivo research, the efficacy and the safety of the drug are determined. In this process, the candidate drug identified in the previous process is refined and tested in a laboratory. This process involves formulation development where solubility, frequency, and stability of the formula, are assessed. The best dosage and administration route is determined. Also, it governs the side and adverse effects of a drug. It compares the effectiveness of the focused drug as compared to the other similar drugs available. The pre-clinical trials are performed by scientists with unlimited dosages. This process is a Pharmacokinetic (PK) process that rates the practices in which the new drugs will affect the human body. This considers where the disease or problem occurs, such as the sinuses or tumor cells. Thus, this process assesses drug safety regarding Absorption, Distribution, Metabolism, and Excretion (ADME) [15]. The major role of this process is to provide sufficient evidence before clinical trials on human beings in terms of safety and efficacy. Proof of Principle (PoP) are studies that are productive in pre-clinical trials and early safety testing. In Vivo, In Vitro, and Ex Vivo Assays are the three types of studies that are conducted in this process. In Vivo is popularly performed on animals, In Vitro is performed in a lab, in Ex Vivo tissues of the non-living animals are examined. In silico assays are an experiment that is performed using computer simulations. These studies are performed on the non-living organisms and tissue extracts. This assures the safety of usage of drugs in human beings by performing various tests on animals that also help to calculate the appropriate doses to test in humans. It also confirms the adequate quantity of availability of drugs before the clinical trials [16]. These assays are used to test a compound's safety and potentially toxic effects using animal models, alternative models, like Zebrafish, or cell cultures [17]. The drug delivery could be oral or topical which is followed by Formulation Optimization & Improving Bioavailability. Formulation optimizations make certain that the drugs are delivered timely to the appropriate place and in the proper concentration.

3.2.3 Clinical drug development and regulatory approval process

The next process is the clinical research process, where drugs are tested on human bodies to make sure that drugs are safe. In this process, voluntary research is carried out to find answers about the effectiveness, safety, and new ways of using existing therapies. The complexity, implementation issues, and costs involved in the clinical trial may influence trials carried out in this phase. These trails should be safe, effective, cost-efficient, and fulfil the intended purpose. It involves Phases, namely Phase 1, which is a healthy volunteer study where the drug is tested on humans to assess the safety and pharmacokinetics, absorption, metabolic, and elimination effects on the human body and determines the adverse effects. Phases II and III are studies on patient populations, where the drug safety is assessed by delivering standard drugs previously used as treatment. It helps to find out the optimal dosage and schedules. The third phase is performed on a large population, permitting medication labelling and appropriate instructions for drug usage. In this phase, extensive collaboration is required with involved of review boards, followed by full-scale production and final drug approval.

3.2.4 FDA review

The Food and Drug Administration (FDA) review process is conducted once the new drug has been invented and results from clinical trials are obtained. Finally, the final process is regulatory approval when all the clinical trials are accomplished. The regulatory approval is performed before a drug is sold on the market. A national regulatory authority performs the drug's approval [18]. The FDA is a governing body responsible for ensuring public safety against poor-quality drugs and medicines. Initially, a drug developer needs to submit all relevant documents of tests and pre-clinical and clinical research. This FDA then reviews all the submitted documents and decides on the approval of the drug. Furthermore, FDA also executes Post-Market Drug Safety Monitoring and has full rights to add cautions to the dosage and usage information [19]. A new drug may fail in the drug examination conducted by the FDA. There might be multiple reasons for this, including toxicity, effectiveness, PH properties, bioavailability, or inadequate drug performance. Toxicity in a drug can be very harmful to the human body, also, if the evidence is questionable, or poor bioavailability or inefficient functionality of the drug as per the expected performance may cause its rejection by the FDA.

3.2.5 Post-market monitoring

The next process is post-market monitoring, which is conducted through the FDA Adverse Event Reporting System (FAERS). This reporting system is performed post-marketing, which involves safety surveillance. The outcome of this surveillance is reported by the manufacturers, health professionals, and consumers of the approved drugs.

In the current era, the research and development process has been revolutionized drastically as compared to the traditional approach. This includes faster delivery to the patients, reduced development costs, improved vision and insights, and more efficient decision-making. The major pillars of excellence in the development process involve the patients and healthcare professionals and involve enhanced design thinking process starting from trial designing to trial execution, approvals, and launch. The redesigning process, which includes both cross-functionally and within the research and development teams that, will further enhance development speed and build on clearer research insights. The cross-functionality could be seen in animal toxicology, biomarkers, chemistry, manufacturing and control (CMC), discovery sciences, drug metabolism and pharmacokinetics, protein, and cell engineering. The digitization of the whole process involves recent technologies that will make the system autonomous and save time for the recurring process with the generation of a new vision and insightful data. The enhanced and intelligent decision-making process which incorporates various analytical tools and predictive modeling. This will eventually enhance the quality and speed of the decision-making process. The incorporation of agile methodologies and portfolio management makes the whole working model of drug discovery optimized. The smooth transfer of materials and information within and across the teams, clear visibility of the critical path, incorporation of risk tolerance, and prioritizing the key activities will improve the drug development process. Table 3.1 presents the life cycle of the drug discovery process.

Table 3.1 Lifecycle of drug discovery process

Discovery and development process	**Preclinical research process**	**Clinical drug development and regulatory approval process**	**FDA review**	**Post-market monitoring**
• Target Identification • Target validation • Hit discovery • Assay development and screening • High Throughput Screening • Hit to Lead • Lead Optimization	• In Vivo • In Vitro • Ex Vivo Assays • ADME • Proof of Principle (PoP) • Drug Delivery • Bioavailability	• Phase I • Phase II and Phase III • Safety and Efficacy • Pharmacokinetic	• FDA approval • Drug Registration	• FAERS

3.3 Pharmaceutical supply chain management and its challenges

The pharmaceutical supply chain is a complex process. The pharmaceutical should ensure that patients get the appropriate medical treatment in a timely manner. The pharmaceutical supply chain defines the protocols in which the prescribed medicines are manufactured and delivered to patients. If the distribution of drugs is conducted incorrectly, it will eventually lower the company's reputation as well as customer satisfaction. The stakeholders involved are manufacturers, wholesale distributors, etc. [20]. The pharmaceutical supply chain includes innovating and developing new products, manufacturing, packaging, and distributing to wholesalers, retailers, pharmacies, and directly to the patient. The pharmaceutical companies do not work on an on-demand basis. Rather the drugs are produced and stored in the warehouse. This may decrease the cost-effectiveness of companies. Also, the pharmaceutical supply chain involves features that are different from other supply chain management products, such as urgency, importance, storage, transportation, and regulation. Figure 3.2 shows the overview of the pharmaceutical supply chain. The major components of the Pharmaceutical Supply Chain include primary manufacturing (researchers), secondary manufacturing, market warehouses distribution centers, wholesalers, and retailers/hospitals. Active Ingredients (AI) are produced in the primary manufacturing site, which involves chemical synthesis to develop complex molecules [21]. Secondary manufacturers are responsible for taking the active ingredient produced from the primary manufacturing site. Also, this stage adds excipient materials and produces the final products. This could be clearly understood by the example of pill manufacturing and marketing. It involves granulation, compression, coating, quality control, and packaging. The secondary manufacturing locations are often geographically separate from the primary manufacturing locations. There are often many more secondary manufacturing sites than primary ones, serving local or regional markets.

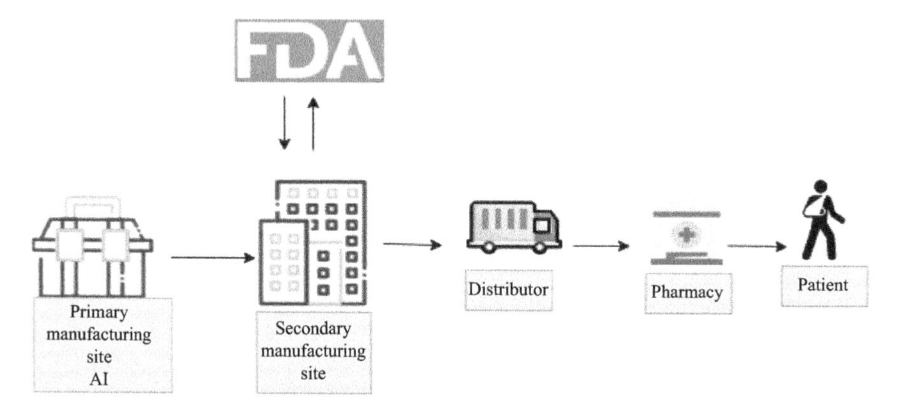

Figure 3.2 Overview of pharmaceutical supply chain

The AI manufacturing sites are responsible for delivering the raw materials to the regulatory agency FDA for approval. The secondary manufacturer sites package the drugs and send them to the distributor. Based on product demand, the distributor is responsible for transferring packaged drugs to pharmacies. Finally, the pharmacy distributes the drug to patients based on prescription. The UPS or FedEx usually supports the transportation task throughout the supply chain [22].

The major challenges in supply chain management include demand management, which requires forecasting the demand of a particular drug based on some geographical location using historical data, market intelligence, etc. The next major challenge is inventory management which includes imposing the demand on the warehouse/distribution center. Other challenges include secondary production planning and scheduling, primary manufacturing campaign planning, and AI inventory management. Relatively large stocks of AI must be held to ensure good service levels and smooth operation at these processes' interfaces. Furthermore, in the case of the pharmaceutical supply chain, the main challenge includes supply chain visibility, drug counterfeiting, cold-chain shipping, etc. These issues lead to significant cost overhead from a consumer perspective. Also, the issues with the pharmaceutical supply chain include maintaining the drug quality and on-time delivery. Furthermore, few drugs have certain temperature boundation that are to be met during transportation. Due to the complex supply chain management of the pharmaceutical industry, drug counterfeiting is a prominent phenomenon that causes fraudulent medicines to pass the trials and verification [23,24].

Fundamental principles of supply chain management are as follows:

1. Involvement of several entities in moving the goods from the place of origin to the final consumer.
2. Several organizations understand how to manage the activities of the supply chain to deliver the product to the end user, hence the majority of firms are limited to their own internal supply chain management operations.

There are two types of flow in supply chain management one is information flow physical flow. Information flow refers to the coordination of regular information regarding the transit of commodities among different supply chain participants whereas the most obvious aspect of supply chain management is physical flow, which involves the movement and transfer of commodities from one location to another.

3.4 Supplier consortiums and preventive maintenance

The pharmaceutical industry involves various stakeholders such as manufacturers, wholesale distributors, and retailers. These stakeholders must agree to a mutual agreement in terms of quality, proper approval strategies, pharmacy, drugs, manufacturing dates, and expiration dates to maintain the supply chain management without harming the safety of the patients. Thus, the aim is to develop a consensus that provides protocols and strategies for effectively implementing the pharmaceutical supply chain to smooth the team-based process of a complex supply chain.

The supplier consortium should sign an agreement regarding quality that takes care of the stakeholders following Current Good Manufacturing Practice regulations (CGMPs) imposed by the FDA [25]. The labs in which drugs are developed are granted a patent after thorough scrutiny. The labs sign a license agreement with the drug company for marketing the drugs and supplying them to the distributors. This agreement between a lab and drug company is performed because the labs do not have big plants to produce many drugs that meet the consumers' demand. Also, there are other agreements such as product supply agreements and research and development agreements. The issues with such agreements are poor regulatory compliance. Regular maintenance is required to pursue research and development. The laboratories should be well equipped to meet the current technology and standards. Also, inefficiencies in operations such as missing expiry dates. This is due to the inefficient conventional approach to document storage. Thus, such agreements should not disclose critical information and maintain confidentiality, authorization, integrity, and privacy preservation [26].

The Pharmaceutical Industry that develops, manufactures, and distributes, drugs are required to be approved for use as medications. These industries follow certain protocols that standardize, patent protection, testing, which ensures the safety, feasibility, and marketing of pharmaceuticals. The various agreements are briefly described as follows.

3.4.1 Quality agreements

The quality agreements are the contracts signed by parties that are participating in medicine manufacturing. These agreements define how each party will fulfill Current Good Manufacturing Practices (CGMPs). CGMPs are the guidelines being imposed by the FDA. The agreements are signed between supplier and vendor for quality assurance of the materials that are been provided to the drug manufacturing plant. The FDA is the governing body that promotes parties that are involved in an agreement for manufacturing to follow quality management best practices. These guidelines are built upon the principles outlined by the International Council for Harmonisation (ICH). This describes the support contract for manufacturing.

3.4.2 License agreements

The License agreement is between two parties, such as a drug company and a pharmaceutical lab. The lab creates new drugs and requests a patent. The drug company is given the license by the laboratory to produce these drugs. These kinds of agreements are usually common in such marketplaces where the lab does not have the resources to produce drugs on a large scale. Thus, the labs grant the license to the big firms that manufacture the drugs that have the required resources available in large quantities.

3.4.3 Product supply agreements

The product supply agreement is an agreement that has been signed by parties such as supplier and buyer. This contract is signed to provide and purchase medication and other pharmaceutical items. In this agreement, the proper guidelines to be followed by both parties are mentioned, in which both have to agree. Also, it mentions all repercussions that will occur on violations of the terms and conditions.

3.4.4 Research and development agreements

In research and development agreements, the pharmaceutical industries enter into collaborative partnership agreements. In this, the parties involved in the contract collaborate to enhance the research and development work that will be carried out. The agreement is for technology among pharma industries and other scientific technology suppliers, including biochemical factories that will supply equipment that is required in the labs to conduct research.

There are various challenges faced while drafting such contracts that are poor regulatory compliance. The various pharmaceutical and biotech industries must maintain to carry out research and development. As with time, it may happen that the industries lack the scientific equipment required to verify that the parties are fulfilling. As the worldwide standards are changing frequently with time. Also, the legal agreement differs for each location or nation. These all variations may make the drug clinical trials process difficult in terms of management and budgeting. The other challenges include inefficiencies in operations as the lack of modern and autonomous technologies may lead to missing expiry dates. The inefficiency can be in terms of sending the wrong documentation and paperwork for approvals. This inefficiency occurs due to the lack of an appropriate decentralized approach for storing the documents. The clauses that are mentioned in the contract include the following:

Recitals are clauses that are non-popular but highly efficient. They serve as the backbone of the whole contract. This clause is required to create a model of the entire agreement; it is established before writing the contract in detail. This model should clearly mention the parties involved in the agreement. Also, it clarifies the purpose and intentions of the involved parties. The parties involved in the agreement are referred to as contract signatories. In case of any dispute, the contract signatories are compelled by the contract, not any third party. The next clause is the scope and purpose of the agreement, which is the most important section of the entire agreement. The clause covers significant details such as qualification, calibration, and maintenance of manufacturing equipment. It also mentions the validation of computer systems, analytical procedures, and manufacturing processes. Furthermore, these class specifications are used to pass or fail analytical tests, supply handling, lab records management, and deviation logging. The next clause is confidentiality, which is required to be maintained for the information to be shared between the parties. This obligation binds the parties involved that if the information has been leaked by one of the parties, then the other party can take the accountable party to the proper authorities to take further actions. This clause ensures that both the parties involved understand the sensitivity of the information being shared. The other clauses are a liability that binds the contractual parties, their representatives, agents, and other subordinates. The next clause defines the limits of the parties involved in terms of Intellectual Property Rights. This defines the ownership of IP rights to provide the license to the other parties involved in this contract. It suggests that if one party gives IP rights to the other party to use their IP rights, this does not result in the other party being authorized of the IP rights. It is all in the hands of the authorized party that owns the IP rights to decide the limits of the use of their IP. The next important clause is the assignment clause, and this allows one party to transfer ownership and allocate agreement duties and rights to another party. This means the assignment clause allows the reassignment of an agreement to a different party. The next significant clause is the dispute resolution clause which takes care of any disputes that may arise between the contractual party involved. This obligation helps to settle down any disputes that may happen between the involved parties throughout the contract duration. The first requirement is the settlement clause must be well written. This is essential because in case of any disagreement between the parties involved in the contract, the relationship should not fail. The contract should be well-drafted to resolve the conflict amicably and quickly between the parties. The last and the most significant clause is the termination clause which is the concluding draft of the agreement. Every contract should terminate in the default case at the end of the term. This should be strategically planned and drafted, which should be an effective escape. The terms and conditions should be well drafted for the smooth termination of the contract. The termination process is activated upon achieving certain circumstances. The condition for the termination should be well drafted in the clause without any ambiguity. Also, it is essential to clearly define the termination time of the contract in this clause.

3.5 Clinical trails

Clinical research is a medical study conducted on humans. It is of two types: observational studies and clinical trials [27]. Clinical trials aim to test a pharmaceutical, surgical, or behavioral intervention using human research [28]. Researchers use them to establish whether a novel therapy, including a mediation, diet, or medical device is effective and safe in humans. Generally, a clinical trial is conducted to evaluate if a new therapy is more effective or has fewer side effects than the present medication [29]. Other clinical studies examine early illness detection methods, often before symptoms' onset. Still, others investigate methods of disease prevention [30]. A clinical trial may also investigate improving the quality of life for those with a life-threatening illness or a chronic health condition. Sometimes, clinical trials examine the role of support groups [31,32].

Clinical trials progress through four steps to test a medication, determine the optimal dose, and examine adverse effects [32]. If, following the first three stages, researchers determine that medicine or other intervention is safe and effective, the FDA authorizes its clinical use and proceeds to monitor its impacts. Generally, pharmacological clinical studies are defined by their phase [33]. Biomedical clinical trials of investigational medicine, therapy, medical technology, or behavioral intervention may involve four phases [28]. The FDA typically needs Phase I, Phase II, and Phase III studies to be undertaken before approving a medicine for use.

- A Phase I study evaluates an experimental medication's safety and side effects on a small sample of patients (20–80), who are typically healthy, to determine the optimal dose [28].
- Phase II trials enroll additional participants (100–300). Phase I has an emphasis on safety, whereas Phase II places emphasis on efficacy. This phase attempts to collect preliminary data on the drug's efficacy in patients with a specific ailment or condition. Additionally, these trials continue to investigate the safety, including short-term negative effects. This period may persist for a number of years [28].
- A Phase III trial collects further data on the medicine's safety and efficacy by examining new demographics, doses, and drug combinations. Typically, the subjects range from several hundred to around three thousand individuals. The experimental medication or gadget will be approved if the FDA agrees that the trial results are positive [28].
- Phase IV trials for medicines or medical devices are conducted following FDA approval. The effectiveness and safety of a gadget or medicine are monitored in broad, varied groups. Occasionally, the adverse effects of medicine may not become apparent until more individuals have taken it for a longer time [28].

Clinical trials evaluate medical interventions, such as medications, cells and other health fluids, surgical operations, radiological procedures, equipment, behavioral therapies, and preventative care [34].

Before they can begin, clinical studies must be meticulously prepared, evaluated, and finished, as well as approved. Clinical trials are accessible to participants of all ages, especially children [35].

Instances of clinical trial objectives include evaluating the relative safety and efficacy of a drug or medical device [36]:

- Regarding a certain type of patient
- At varied doses
- Regarding a fresh indication
- The efficacy of treatment compared to conventional treatment for a certain ailment is evaluated
- Evaluation of the studied medicine or device in comparison to two or more approved/common therapies for the condition

Each research study comes with its own set of rewards and dangers. People need to get all of their questions addressed before they can make a decision about participating in one. Some individuals bring a buddy with them, make notes, or film their sessions with the clinical team in order to assist them in remembering the answers and thinking of more questions [22]. Every clinical study comes with its own set of advantages and potential drawbacks. The majority of clinical studies, though (those that are not in phase 0), offer some of the same possible benefits as the following:

1. By contributing to the furtherance of healthcare research, the user may be able to assist other people who suffer from the same condition.
2. The patient receives a therapy that is not offered anywhere else outside of the clinical study. This treatment may be less risky or more effective than the ones that are now available.
3. Due to clinical trials, the number of therapy options that are available to the patient may expand.
4. If the patients take a more active part in managing their own health care, the patient could have a greater sense of control.
5. It is possible that the patient will have to visit the healthcare team more frequently so that they can monitor the condition and check for any adverse effects the new therapy may cause.
6. During the course of the trial, certain sponsors of the research may pay for all or a portion of the patient's medical treatment as well as any additional expenditures. (This is not always the case with clinical studies.)

Participating in a clinical study comes with a number of potential dangers, including the following [23]:

1. It is possible that the novel treatment has some undetected adverse effects or other hazards and that these risks might be higher than those associated with the traditional therapies.
2. Even if the new therapy helps other people, it is possible that it will not work for the patient.
3. The patient could require more tests or visits to the doctor, both of which could take up more time and force the patient to travel further.
4. If the patient participates in a randomized clinical trial, it is possible that the patient will not have a say in the treatment he ultimately receives. If the research is blinded, neither patient nor the attending physician will be able to determine which treatment he is receiving. (The research team conducting the clinical study will have access to this information in order to ensure safety at all times.)
5. It is possible that insurers will not pay for all of the expenditures associated with the clinical study. They often do take care of the expenditures associated with what is considered to be regular medical treatment. Before the person makes the decision to participate in the clinical study, the person should consult both the health insurance provider and a member of the research team.

Issues that frequently arise during clinical trials:

The majority of people have some reservations about participating in a clinical trial due to the fact that they are unsure of what exactly the benefits of doing so will be for them. Collect as much data so that the researcher can confidently make the decision that is best for the patient [21].

(a) **Will there be risks?**

Yes, there is a possibility of harm in every clinical experiment. However, there is a danger associated with every medical examination, treatment, or surgery. Due to the increase in unknown factors, there is a possibility that the risk will be greater during a clinical trial. This is particularly true in clinical trials in phases I and II when a smaller number of participants get therapy and the treatment is evaluated.

(b) **Will I get a placebo?**

A placebo is a counterfeit pill or therapy that is used in some kinds of clinical studies. Its purpose is to assist researchers in verifying that the observed effects are the consequence of the newly developed treatment or medicine. The term "sugar pill" can also be used to refer to a placebo pill. In clinical studies, the use of placebos as the only therapy is extremely uncommon unless there is no proven effective treatment. In the vast majority of clinical studies for cancer, placebos are not used unless they are also being administered in conjunction with active treatment. It is immoral to provide someone with a placebo rather than therapy that has been demonstrated to be effective.

(c) **Will I be informed as to which group I belong to? Will my physician be aware?**

Each research is unique in its own way. A blinded study is one in which the patient is not informed as to the treatment they are receiving. Neither the participant nor the treating physician is aware of which therapy is being administered during a double-blind trial. It might be challenging to go into a situation without knowing what to expect. If there is a significant medical cause (such as a potentially adverse response to medication), the physician can always find out which group the person belongs to, but, doing so may result withdrawn of the person from the research. Blinding decreases the likelihood that the physician or patient's preconceived notions about the new therapy would influence their assessment of the patient's response or the potential for adverse consequences.

(d) **Will I be a "guinea pig?"**

It is true that the objective of a clinical trial is to address a medical question; nevertheless, the individual will not be used as a test subject in this experiment. It is possible that individuals who participate in clinical trials will be required to perform additional tasks or undergo specific examinations as a result of the study.

However, this does not imply that the individual will not receive good care while the individuals are participating in the research. In point of fact, the vast majority of participants in clinical studies express gratitude for the additional attention they receive from their health care team.

According to a number of studies, patients afflicted with cancer who reported feeling confident in the information they had before taking part in a clinical trial experienced a lower level of post-study regret.

(e) **Will there be a guarantee that my information will remain confidential?**

The identity and any relevant health reports of the user will be shielded from public view to the greatest extent practicable. The primary healthcare team of the user absolutely needs these data in order to provide with the best possible treatment; in fact, they would need them even if users were not participating in a research study.

The information that is essential for the clinical study, such as the results of the tests, is recorded on specialized forms and entered into computer systems. This information is only disclosed to those who are responsible for analyzing the outcomes of the research. The information will be identified only by a number or code; neither the forms nor the research database will contain the individual's name. It is possible that members of the study team or representatives from the FDA will need to review the records in order to validate the information that they have been provided with. However, they will not have access to any of person's personal information, nor will it be utilized in any public reports of the findings of clinical trials.

(f) **Medicare coverage for clinical trials**

In the Medicare, the government pays for a significant portion of the usual medical expenses for cancer patients who are participating in clinical studies that have been approved. In most cases, Medicare will pay for cancer treatment if it falls under one of the following categories:

- A diagnostic or therapeutic clinical study for cancer, or
- A clinical study that is either sponsored or supported by the National Cancer Institute (NCI), one of the NCI-Designated Cancer Centers, one of the NCI-Sponsored Clinical Trials Cooperative Groups, or another government organization that is responsible for funding cancer research.

(g) **What steps might the researcher takes to acquire further information regarding prices**

Collect as much information as about the clinical study, and then get in touch with the insurance carrier to inquire about how payment would be handled. Because many insurers evaluate claims on an individual basis, it is possible that they would not be able to provide the patient with a straightforward yes or no response to the question. In addition to this, they will check to see if the oncologists who are providing the majority of cancer treatment are "in network."

In the event that the health insurance will not cover some aspects of the clinical study, discuss the alternatives with either the primary care physician or the research coordinator. There is a possibility that sponsors will agree to cover part of the expenditures that the insurance would not.

3.5.1 Need for traceability in clinical trials

Traceability indicates that every unique item of clinical data may be tracked throughout the complete software flow of all applicable application programs, from data acquisition to final submission papers. Again the raw data, documents, and files at any step in the system may be inspected for accuracy and completeness [37]. Conventionally, information is made useable by collecting raw data and entering it into a data management device, which produces a raw data repository for use by statisticians and programmers [38]. Code is created to construct derived variables that are saved with the raw variables to give a database for reporting in support of a submission. Then, another code is generated to generate the tables, lists, and figures that will be utilized in the initial instance for the Central Serous Retinopathy (CSR). A drawback of this method is that these summary tables may include crucial information that is never stored as data points and is only traceable through a cut-and-paste operation or an interface across environments [39]. The output tables and documents provide additional locations for storing the same data. This procedure is repeated when data is extracted from the CSR and incorporated into other regulatory papers, such as reporting and disclosure reports, investigator brochures, clinical trial protocols, and submission forms [40]. The singular truth becomes diluted and dispersed over several publications, forms, and locations.

Traceability in clinical trials:

(a) *Data-point traceability*: During a clinical trial, numerous data points are generated utilizing various procedures and computer systems. Gathering raw data and loading it into an information management tool, which is subsequently removed to give a raw data repository for statisticians and developers, is a conventional method for transforming raw data into useable information [41]. To give a reporting database to complement the submission, a Code is generated to construct derived factors that are placed with the raw variables. Code is then produced to generate the tables, lists, and figures (TLFs) that will be used for the initial case of CSRs [42]. As information is taken from the CSR and incorporated into other regulatory papers, such as Investigator's Brochures, Public Disclosure Reports, Clinical Trial Regulations, and Submissions forms, the same procedure is repeated [43].

(b) *Metadata traceability*: It provides the documentation necessary to clearly describe information in the Study Data Tabulation Model (SDTM) datasets, as well as the procedures and methods utilized to obtain the analysis result [41]. Metadata governs the maintenance of standards and the data and information's traceability [44]. If metadata is utilized to store and monitor data, then it is possible to trace assertions in documents back to the original data. This approach facilitates the searchability of relevant data [45,46].

Traceability is vital for supporting clinical research analytic findings since the validity of the study results depends on the source data and the quality and repeatability of the techniques employed [46]. The traceability of a particular information point can begin to use this kind of surroundings. When a query is made, it is possible to find out the data that was utilized in the history of the implementation of such information [47]. Finding the only accurate account of events promptly and with complete assurance. When clinical data services are centralized, they offer sponsors a stable environment to conduct their clinical trial filing processes. Information points during the time are only kept once, advanced to keep in step with the expansion of the specification and are always available. Traceability of information across time is controlled from the first day of a product's lifespan. When people have questions regarding the product, the replies are controlled through the informational trail that has been produced. This ensures that the sponsors always have access to the same version of the truth.

When it comes to clinical information management, all life science companies are up against the same obstacles [48]. It is important to consider the fundamental principles of how data, information, knowledge, and wisdom are generated and how these can be traced throughout a product's lifecycle to be more efficient in managing clinical knowledge and information. This will allow the process to be more efficient in managing patient knowledge and information [49]. In addition, it is essential to have a solid understanding of the information in place to handle the label of a product, as well as the rate and precision with which that information can be accessed. This can be accomplished by being familiar with the information flow across and within functions [49].

The final measures that determine how well we are doing in terms of the application relate to the amount of time it takes to get approval as well as the quality of the proposal. In essence, if we can supply data of such high quality that it is unnecessary to ask a question, then our reaction times will be kept to a minimum [50]. This is a genuine assessment of the quality of the information that supports the assertions that are made in the label [51].

3.5.2 Revolutionizing clinical trials using blockchain technology

Blockchain is a technology to significantly affects clinical trial supply chains by enhancing the traceability of pharmaceuticals through Active Pharmaceutical Ingredient (API) to the participant and allowing the HIPAA-compliant collection of patient-level data [37,52]. This is accomplished by having patients and other network participants record information to the blockchain, which subsequently transfers the data to the proper system and organizations with permission to read the data [52]. The auditable, immutable information may be used to construct a continuous record of a patient's health condition. Blockchain is essentially a decentralized and encrypted version of a massive, shared database [53]. It is generally recognized as an essential component of the digital asset infrastructure, whose dependability is key to assure. It is responsible for the transparency and traceability that enable a wide variety of services [54]. In this context, the developing and promising technology known as blockchain has the potential to deliver a strong basis for transparency of the enrolling phase to all of the stakeholders of a clinical trial [55].

This is particularly important when it comes to the context of getting participant consent [56]. The following are the three key functional concepts that underlie this technology and have the potential to play a significant role:

- Unfalsifiable timestamping data, or evidence of the presence of any data. These data, when saved, are verifiable and unchangeable via a robust cryptographic mechanism. Furthermore, this existing evidence may be verified on a public website. This transparency is beneficial for all stakeholders [57].
- A smart contract is a contract that is programmatically built and binds each protocol modification requiring renewed patient permission [58].

- The protocol's decentralized structure offers the patient or, more broadly, the patient community power over their agreement and its withdrawal. The end-to-end connection generates a peer-to-peer network between patients and researchers [59].

The current implementation demonstrates the first concept. Idealistically, we must develop a patient verification system that does not depend on any trial participant to take advantage of the distributed and trustless characteristics of the Blockchain network [60]. The secure timestamping capabilities of blockchain can directly preclude an a posteriori rebuilding of endpoints or results in clinical studies in a larger environment.

The blockchain plays in two scenarios:

At the clinical trial level: Blockchain technology can serve as a SafeGuard for clinical trials' complicated and extensive cast of characters [61]. In practice, the proof of existence for consent is timestamped and stored in Blockchains, allowing clinical research stakeholders, such as funders, investigators, and IRBs, that can be numerous in multi-center medical examinations, to share consent and re-consent related data in real-time time and to archive and historicize consent sets, which can be paired with each modification of the protocol.

At the patient level: Incorporating "privacy by design" technique and storing all necessary datasets in a safe and transparent manner is a significant step toward enhancing the enrollment phase technique. Obviously, there are techniques to archive data with a certain amount of security. Prior to the invention of blockchain, distributed ledger technologies enabled such security. However, the breakthrough in blockchain's validation protocol, known as "proof-of-work" in the Bitcoin network, enabled the design of an open, inclusive system in which peers contribute to the internet backbone effort to appropriate material, also known as "mining." In addition, utilizing methods to acquire informed consent safely and transparently, respecting participant rights, and empowering them might increase the enrollment rate. The rate of participation in clinical trials remains relatively low [62]. A rigorous comparison of alternative enrolment approaches revealed, among other factors, that gaining patient permission in an open and safe environment was the most effective method for increasing the enrolment rate.

The findings of a recent research that was carried out to better understand and raise awareness of the challenge of patient enrollment in clinical trials were disastrous. Only 16% of patients were genuinely aware of pertinent clinical trial possibilities when they were exploring treatment options, according to research that looked at how many participants were knew about ongoing and accessible clinical trials. That indicates that almost 85% of patients were unaware of the importance of clinical trials or the fact they would have benefitted from the treatments the trials offered. It is amazing and should be made known to the public that a great deal of time and money is spent on researching pharmaceuticals, but little any is spent on locating individuals for clinical trials. In a variety of ways, blockchain technology has the potential to directly boost the number and caliber of patients enrolled in clinical trials. Individual patients could be able to record their medical information on this kind of distributed ledger using anonymous means, making it accessible to trial recruiters who could then get in touch with the patients if their information qualifies them for the clinical trial. Additionally, it could make doctor–patient communication during the experiment more efficient. Blockchain technology may be used to create smart contracts that support transparency and traceability in the conduct of clinical trials and can offer rewards for patient engagement and data sharing.

In the clinical prescription supply chain, an underlying analysis finds three problems that blockchain can solve:

- The network between a clinical trial sponsor, a study subject, and a location is lengthy and incorporates numerous IT systems. In a world where all parties are connected via a blockchain, it would have been possible to leverage encryption and security systems so that members could obtain confirmation of the invoice of the product without getting access to secured patient information, thereby enabling the validation of patient identity.
- Before initiating a clinical trial, the funder is required to submit the study's parameters to the relevant regulatory authorities and numerous Ethical Committees (ECs) at the trial locations. At the research completion, the sponsor's regulatory affairs department verifies that all requirements have been met before accepting and submitting the findings to the appropriate authorities for approval. Sponsors and regulators confront the difficulty of ensuring the integrity of the data and establishing standard rules for this procedure. The introduction of blockchain technology can facilitate the implementation of such standards, as the validity and traceability of transactions are fundamental to the platform.
- Blockchain technology can help the complete clinical trial process by introducing smart contracts. The procedure milestones can function as checkpoints for the intelligent contract. The procedure will only continue beyond that stage when all necessary tasks have been accomplished and are accurate.

Blockchain technology offers several benefits in terms of data security and the potential to connect diverse systems utilized by manufacturers, CROs, and research sites. This technique provides the advantages of centralization without requiring all data to be stored in a single location, making it less susceptible to external and internal threats [63]. A combination of digitalization and blockchain technology plays a vital role in preserving the integrity of clinical trials and overcoming obstacles. Incorporating the latest technical expertise gives a level of certainty that is lacking in the market today. Integrating blockchain technology into clinical research has a wealth of rewarding advantages, from data monitoring and sharing to the required openness and patient privacy issues. The implementation of this technology will have a monumental and urgent worldwide impact. Although the development of blockchain technology and clinical research studies is still in its inception, experts are always working together to identify new methods to reduce costs and save time for a variety of applications in the clinical trial environment. In the end, their efforts will lead to more affordable medications for everyone. Recent studies have shown how difficult it is to reproduce clinical trials. The outcomes of duplicated studies may not always match those of the original clinical study since many experiments are not repeated. This problem is partially caused by data entry errors voluntary or not by the numerous actors that manage the data. A Blockchain is a fantastic tool for addressing this issue. In fact, the information kept there is organized strictly according to chronological order and cannot be changed. This strengthens the clinical trial's data and metadata's consistency. Data, protocol, and previous protocols are all stored in memory and checked as new data is added. This significantly reduces the possibility of data entry mistakes or even of inadequately reporting unfavorable events.

3.6 Pharmaceutical supply chain management using blockchain technology

Supply chains are currently becoming highly complex in structure, challenging in terms of tasks and diverse stakeholders. Several businesses do not have an integrated view of the entire supply chain. Several global companies have developed their own identities and platforms to retain global operational exposure and have the ability to direct their suppliers. They have to focus on centralized administrative or intermediary bodies. This low transparency causes many problems and difficulties in the supply chain process in terms of confidentiality, traceability, authentication, and verification method. Blockchain is a revolutionary computer technology that can support many possible operations and supply chain-related applications. It is important to note that blockchain is well suited to the complexities of supply chains. Blockchain technologies can also contribute to the domain of operations and the supply chain. Thus, blockchain technology provides a more robust and efficient supply chain system in the pharmaceutical sector.

The global supply chain of medical goods and services is essential to the healthcare sector. Several supply chain management-related problems are resolved by the use of blockchain technology in the healthcare sector. RFID and EPCIS are only two of the technologies being used to safeguard the medication supply chain (electronic product code information service). The use of technology helps guarantee that the active chemicals used to make drugs adhere to the highest medical requirements. Blockchain technology may also be used to trace the origin of counterfeit medicines and gadgets. For the healthcare sector, several businesses are creating blockchain-based solutions to address these problems. A US-based business called Chronicled created a platform to trace the transfer of medical supplies using blockchain and the Internet of Things (IoT). Manufacturers, pharmacies, and distributors may trace the flow of medical goods in real time with the help of the Chronicled company, guaranteeing that only authorized people have access to them.

Using blockchain technology in the pharmaceutical supply chain system has the potential to improve supply chain transparency, traceability, and security, as well as reduce costs. The blockchain-based supply chain system maintains the trust of the distributors, retailers, and customers by allowing them to track products from origin to reception. It helps all the supply chain members verify the authenticity of the product or the material and prevents fraud. Blockchain technology stores the product or material-related information, such as date, location, quality, unique identification number, and certification, in the form of a blockchain structure. It strengthens the traceability procedure and lowers the losses from counterfeit or gray market trading. Furthermore, the blockchain-based smart contract functions may be utilized as an agreement between the stakeholders to automatically execute the terms of the agreement when conditions are satisfied on both sides. In addition, the implementation of smart contract functions increases the accuracy of the tracking and monitoring module. It tracks the location and status of the drugs in the supply chain system and combats issues like counterfeit drugs, compliance violations, delays, and waste. Also, immediate actions can be taken during an emergency, like in the case of product recalls, and the ledger audit trial ensures regulatory compliance. Thus, combining blockchain technology and supply chain system automates production, transportation, and quality control operations tracking.

Many researchers, academicians, and developers propose the blockchain-based supply chain architecture for pharma supply chain systems to provide transparency, security, and tracking features. For example, Kumar *et al.* [64] designed a Quick Response (QR)-based secure blockchain system to provide complete traceability of medical products from the manufacturer to customers. The proposed system uses the blockchain's advanced features to identify counterfeit drugs and prevent patient harm and even death. Konapure *et al.* [65] provide a smart contract-based architecture for pharma supply chain systems to ensure efficient product tracing. The smart contract eliminates the requirements of a third party or mediator and gives a secure and immutable transaction history. Saindave *et al.* [66] developed an improved supply chain of drugs system using blockchain technology that reduces counterfeits. The developed system ensures transparency among actors and provides security, authentication, and privacy features. Similarly, Kumar *et al.* [67] analyze the cracking down of the fake drug industry using blockchain technology. The authors proposed a quantitative analysis model using blockchain to handle the sensitive information related to drugs and solve the counterfeit problem permanently.

The supply chain travels through each of the different checkpoints involved in product assembly and distribution. Today's supply chains might include hundreds of steps and cover a wide range of geological regions.

Since there are data misfortunes and roadblocks at every step of the supply chain, it is challenging to adhere to a fool proof framework. Furthermore, if the pharmaceuticals industry's supply chain is taken into account, the security of the process flow becomes a crucial role. A fake drug might perhaps include more dangerous ingredients, such as floor wax or boric corrosive. The presence of rodent poison is occasionally observed. These can produce hypersensitivity conditions that may really inflict real harm or even death. Due to its innovative qualities that provide effective solutions for the already acknowledged vulnerabilities, Blockchain technology has been developing fascinating study areas to improve the traceability and security of the supply chain. Blockchain may be summed up as a communicated knowledge base that is shared and agreed upon by a dispersed organization, sometimes referred to as a peer-to-peer network, in simple words.

A component becomes immutable when it is added to the blockchain, transforming it into a changeless record of previous transactions and enhancing both performance and security. To sum up, the characteristics of Blockchain that make it the appropriate answer are:

(a) Immutability
(b) Transparency
(c) Information verification
(d) Secured by cryptography

The two primary services offered by blockchain technology are Ethereum and Hyperledger. These provide context for understanding the accountability problem. Whenever a product exchanges hands, the transaction is documented and stored on the blockchain for verification and trustworthiness. The trust will initially be built through collecting organizations. They will test the raw materials given to them by the provider in a lab before starting the production cycle and doing further testing. The primary goal of this paper is to present the comprehensive, implementable architecture of the supply chain that is supported by both blockchain services. Both techniques will be contrasted based on a number of factors that are important throughout the whole cycle [68–70].

For example, Figure 3.3 depicts the use case of a blockchain-based pharmaceutical supply chain system to track the delivery of medical products using the blockchain structure and smart contracts. It utilized a two-layered architecture based on blockchain and pharmaceutical supply chain systems to store the product-related details in the blockchain structure and provides a complete transparent tracking system. The first layer, named the data layer, involves various entities such as researchers, manufacturers, distributors, retailers, and customers. It manages the interaction between entities and generates drug-related data. The interaction between different entities involves trading the pharmaceutical supply chain system products. The second layer that is blockchain layer maintains entity and product-related data in the form of transactions executed during the trading and delivery of

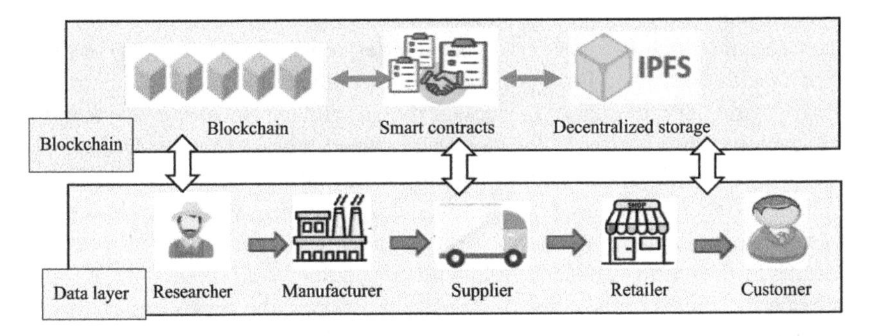

Figure 3.3 Medical product tracking use case based on blockchain technology and pharmaceutical supply chain system

the products. Various smart contract functions such as registration, add product_lot, add location, and reputation are used to provide basic services to the involved entities. The smart contract functions execute automatically on the occurrence of particular events such as trading, delivery, and reputation. It also maintains a decentralized storage system to store the complete details of the product with stakeholders in a decentralized manner using the InterPlanetary File System (IPFS). On execution of the smart contract, the detailed information is stored on the IPFS, and meta details are maintained in the blockchain structure in the form of hash values. Thus, the blockchain layer provides the security features such as confidentiality, privacy, transparency, and auditability.

The blockchain-based product tracking use case involves the following entities:

- **Researcher:** A researcher is the first to execute a smart contract function for trading in the pharmaceutical supply chain system. Researchers produce drugs and take responsibility for monitoring and maintaining the details of the drugs. Then, researchers sell the drugs to the processors.
- **Manufacturer:** The manufacturer buys the drugs from researchers. Then, the manufacturer converts the drugs into the final product. The manufacturer sells the converted product to suppliers.
- **Supplier:** The supplier stores the received products from processors and then sells them to the retailers.
- **Retailer:** The retailer buys the products from distributors and then sells them to the customer per the doctor's prescription. The product is uniquely identified using the identification number of the goods that helps the user to track the provenance data.
- **Customer:** The customer buys and utilizes the final product from retailers. The customer also checks the credibility of the retailer using the reputation system before purchasing the products.

3.7 Conclusions

This book chapter highlights the pharmaceutical supply chain system's significance by covering the drug discovery process, supplier consortiums, clinical trial details, and preventive measures. First, it provides the complete details of the pharmaceutical drug discovery process and discusses the pre-clinical research process, clinical drug development process, review, and monitoring process. Then, the book chapter analyses and evaluates the various challenges involved in the pharmaceutical supply chain system, such as counterfeit drugs, drug traceability, security, and privacy issues. Next, it explains the process of supplier consortium and its preventive measures to deal with the identified challenges, along with the details of various agreements involved in the consortium system. Further, the book chapter discusses the clinical trial process with and without blockchain technology. In the end, it evaluates the concept of blockchain technology in pharmaceutical supply chain systems to tackle the identified challenges by providing the tracking process in a complete decentralized manner with the protection against counterfeit drugs. The book chapter considers the medical product tracking use case to explain the usage of blockchain technology in the domain of pharmaceutical supply chain systems. This use case also explains the concept of the blockchain technology-based smart contract that ensures the various essential features in the pharma supply chain system, such as registration of entities, secure maintenance of records, drug traceability, and attain security features. It also discusses the working of blockchain technology in the supply chain system and provides a basic understanding of implementing blockchain to achieve security, privacy, and traceability features. In future work, we plan to propose a blockchain-based pharma supply chain architecture that resolves the identified challenges and provides an effective solution. We will utilize the various performance evaluation parameters such as latency, throughput, computation time, execution time, and response time, to measure the effectiveness of the blockchain-based supply chain architecture in the pharmaceutical domain.

References

[1] W. G. Chambliss, W. A. Carroll, D. Kennedy, *et al.*, "Role of the pharmacist in preventing distribution of counterfeit medications," *Journal of the American Pharmacists Association*, vol. 52, no. 2, pp. 195–199, 2012.

[2] Z. RJ, "Roles for pharmacy in combating counterfeit drugs," *Journal of the American Pharmacists Association*, vol. 48, pp. e71–e88, 2008.

[3] S. Nakamoto, Bitcoin: A Peer-To-Peer Electronic Cash System. Available: https: //bitcoin.org/bitcoin.pdf, 2008. [Accessed on 13 January 2022]

[4] P. Sharma, R. Jindal, and M. D. Borah, "A review of blockchain-based applications and challenges," *Wireless Personal Communications*, vol. 123, pp. 1201–1043, 2022. DOI: https://doi.org/10.1007/s11277-021-09176-7.

[5] S. Namasudra, G. C. Deka, P. Johri, M. Hosseinpour, and A. H Gandomi, "The revolution of blockchain: state-of-the-art and research challenges," *Archives of Computational Methods in Engineering*, vol. 28, pp. 1497–1515, 2021. DOI: https://doi.org/10.1007/s11831-020-09426-0.

[6] L. S. Sankar, M. Sindhu, and M. Sethumadhavan, "Survey of consensus protocols on blockchain applications," In *Proceeding of the 4th International Conference on Advanced Computing and Communication Systems* (ICACCS), 2017, pp. 1–5.

[7] A. Musamih, K. Salah, R. Jayaraman, *et al.*, "A blockchain-based approach for drug traceability in healthcare supply chain," *IEEE Access*, vol. 9, pp. 9728–9743, 2021. DOI: 10.1109/ACCESS.2021.3049920.

[8] G. Subramanian, A. S. Thampy, N. V. Ugwuoke, and B. Ramnani, "Crypto pharmacy – digital medicine: a mobile application integrated with hybrid blockchain to tackle the issues in pharma supply chain," *IEEE Open Journal of the Computer Society*, vol. 2, pp. 26–37, 2021. DOI: 10.1109/OJCS.2021.3049330.

[9] O. Alkhoori, A. Hassan, O. Almansoori, *et al.*, "Design and implementation of cryptocargo: a blockchain-powered smart shipping container for vaccine distribution," *IEEE Access*, vol. 9, pp. 53786–53803, 2021. DOI: 10.1109/ACCESS.2021.3070911.

[10] O. M. Salo-Ahen, I. Alanko, R. Bhadane, *et al.*, "Molecular dynamics simulations in drug discovery and pharmaceutical development," *Processes*, vol. 9, no. 1, p. 71, 2020.

[11] K. Martinez-Mayorga, A. Madariaga-Mazon, J. L. Medina-Franco, and G. Maggiora, "The impact of chemoinformatics on drug discovery in the pharmaceutical industry," *Expert Opinion on Drug Discovery*, vol. 15, no. 3, pp. 293–306, 2020.

[12] A. G. Atanasov, S. B. Zotchev, V. M. Dirsch, and C. T. Supuran, "Natural products in drug discovery: advances and opportunities," *Nature Reviews Drug Discovery*, vol. 20, no. 3, pp. 200–216, 2021.

[13] M. Muttenthaler, G. F. King, D. J. Adams, and P. F. Alewood, "Trends in peptide drug discovery," *Nature Reviews Drug Discovery*, vol. 20, no. 4, pp. 309–325, 2021.

[14] B. Ding, "Pharma industry 4.0: literature review and research opportunities in sustainable pharmaceutical supply chains," *Process Safety and Environmental Protection*, vol. 119, pp. 115–130, 2018.

[15] D. E. C. Yu, L. F. Razon, and R. R. Tan, "Can global pharmaceutical supply chains scale up sustainably for the covid-19 crisis?," *Resources, Conservation, and Recycling*, vol. 159, p. 104868, 2020.

[16] K. Zhi, B. Raji, A. R. Nookala, *et al.*, "PLGA nanoparticle-based formulations to cross the blood-brain barrier for drug delivery: from R&D to cGMP," *Pharmaceutics*, vol. 13, no. 4, p. 500, 2021.

[17] J. Vamathevan, D. Clark, P. Czodrowski, *et al.*, "Applications of machine learning in drug discovery and development," *Nature Reviews Drug Discovery*, vol. 18, no. 6, pp. 463–477, 2019.

[18] A. Khatoon, "A blockchain-based smart contract system for healthcare management," *Electronics*, vol. 9, no. 1, p. 94, 2020.

[19] I. Siregar and A. A. Nasution, "Component identification the causes of machinery damage in pharmacy company using pareto diagram," In *Proceedings of the IOP Conference Series: Materials Science and Engineering*, vol. 420, no. 1, IOP Publishing, 2018, pp. 012139.

[20] T. B. Lingaiah, Y. A. Belay, and K. Dese, "Developing a desktop application for drug-drug interaction checker ordered for chronic diseases in Ethiopian hospitals pharmacy," *BMC Pharmacology and Toxicology*, vol. 23, no. 1, pp. 1–9, 2022.

[21] H. Yinghua and L. Yi, "Targeted preventive maintenance of pharmaceutical equipment," *Journal of Drug Design and Medicinal Chemistry*, vol. 4, no. 2, pp. 10–15, 2018.

[22] J. S. Mwawaka, "Re-engineering collaborative practice in primary care: Integrating community pharmacy to the clinic; creating a pharmacy referral and quality circle," *Innovations in Pharmacy*, vol. 12, no. 3, p. 12, 2021.

[23] A. Banerjee, "Blockchain in the pharmaceutical and healthcare ecosystem," In *Blockchain in Digital Healthcare, Chapman and Hall/CRC*, 2021, pp. 209–230.

[24] S. Kalarani, K. Raghu, and S. Aakash, "Blockchain-based e-pharmacy to combat counterfeit drug transactions," In *Mobile Computing and Sustainable Informatics*, New York, NY: Springer, pp. 377–390, 2022.

[25] K. Blagec, J. J. Swen, R. Koopmann, *et al.*, "Pharmacogenomics decision support in the u-pgx project: results and advice from clinical implementation across seven European countries," *PLoS One*, vol. 17, no. 6, p. e0268534, 2022.

[26] C.-N. Chen, C.-H. Lai, G.-W. Lu, *et al.*, "Applying simulation optimization to minimize drug inventory costs: a study of a case outpatient pharmacy," *Healthcare*, vol. 10, no. 3, p. 556, 2022.

[27] J. A. DiMasi, G. H. G. Grabowski, and R. W. Hansen, "Innovation in the pharmaceutical industry: new estimates of R&D costs," *Journal of Health Economics*, vol. 47, pp. 20–33, 2016.

[28] Bill and Melinda Gates Foundation, "Clinical trials". Available: https://www.gatesfoundation.org/ideas/media-center/press-releases/2006/04/malaria-clinical-trials-alliance, 2006 [Accessed on 27 July 2022].

[29] US Food and Drug Administration, "Medical devices, premarket clinical studies for investigational device exemption". Available: https://www.ncbi.nlm.nih.gov/pmc/articles/PMC6113340/, 2017 [Accessed on 2 January 2022].

[30] F. Lederle, J. Freischlag, T. Kyriakides, *et al.*, "Outcomes following endovascular vs open repair of abdominal aortic aneurysm: a randomized trial," *JAMA*, vol. 302, no. 14, pp. 1535–1542, 2009.

[31] D. Sessler and P. Imrey, "Clinical research methodology 2: observational clinical research," *Anesthesia and Analgesia*, vol. 121, no. 4, pp. 1043–1051, 2015.

[32] E. Miseta, "Janssen uses geofencing to monitor clinical trial patients," In *Clinical Leader*, Philadelphia, PA: VertMarkets, 2019.

[33] J. Unger, E. Cook, E. Tai, and A. Bleyer, "The role of clinical trial participation in cancer research: barriers, evidence, and strategies," *American Society of Clinical Oncology Educational Book, American Society of Clinical Oncology*, vol. 35, no. 36, pp. 185–198, 2016.

[34] S. Piantadosi, *Clinical Trials: A Methodologic Perspective,* New York, NY: John Wiley & Sons, 2017.

[35] L. M., Friedman, C. D., Furberg, D. L., DeMets, D. M., Reboussin, and C. B. Granger, *Fundamentals of Clinical Trials,* New York, NY: Springer, 2015.

[36] C. Umscheid, D. Margolis, and C. Grossman, "Key concepts of clinical trials: a narrative review," *Postgraduate Medicine*, vol. 123, no. 5, pp. 194–204, 2011.

[37] M. Benchoufi, R. Porcher, and P. Ravaud, "Blockchain protocols in clinical trials: transparency and traceability of consent," *F1000Res*, vol. 23, no. 6,

p. 66. DOI: 10.12688/f1000research.10531.5. PMID: 29167732; PMCID: PMC5676196, 2017.

[38] Applied Clinical Trials, "*Improving the traceability of the clinical trial supply chain,*" vol. 26, no. 12, 2017.

[39] Quanticate CRO Blog, "Traceability – where is your clinical information?". [Accessed on 27 July 2022].

[40] D. Glover and J. Hermans, "Improving the traceability of the clinical trial supply chain," 2017.

[41] Worksure, "Clinical data – importance of traceability in clinical data management". [Accessed on 26 July 2022].

[42] G. White, "Metrological traceability in clinical biochemistry," *Annals of Clinical Biochemistry*, vol. 48, no. 5, pp. 393–409, 2011.

[43] H. Vesper and L. Thienpont, "Traceability in laboratory medicine," *Clinical Chemistry*, vol. 55, no. 6, pp. 1067–1075, 2009.

[44] M. Zozus, J. Bonner, and L. Rock, "Towards data value-level metadata for clinical studies," In *ITCH*, pp. 418–423, 2017.

[45] J. Philips, "Data and metadata traceability in engineering research to improve transparency, accountability, and explainability," In *VLIR/VCWI Biannual Integrity Event*, 2021.

[46] S. Hume, S. Sarnikar, and C. Noteboom, "Enhancing traceability in clinical research data through a metadata framework," *Methods of Information in Medicine*, vol. 59, pp. 75–85, 2020. DOI: 10.1055/s-0040-1714393.

[47] M. Tohen, "Credibility of industry-funded clinical trials," *The International Journal of Neuropsychopharmacology*, vol. 16, no. 8, pp. 1879–1884, 2013.

[48] M. Tohen, "Treatment guidelines in bipolar disorders and the importance of proper clinical trial design," *The International Journal of Neuropsychopharmacology*, vol. 20, no. 2, pp. 95–97, 2017. DOI:10.1093/ijnp/pyx002.

[49] M. Edwards and S. Kochhar, "Ethics of conducting clinical research in an outbreak setting," *Annual Review of Virology*, vol. 7, no. 1, pp. 475–494, 2020. DOI:10.1146/annurev-virology-013120-013123.

[50] P. Span, "When a drug study abruptly ends, volunteers are left to cope," *The New York Times*, pp. 0362–4331 [Accessed on 23 April 2022].

[51] J. D. Santomauro, "An exploration of the global clinical trial ancillary supply chain and the drivers of success during the pre, in, and post phases," (Doctoral dissertation, Temple University. Libraries), 2019.

[52] S. Apte and N. Petrovsky, "Will blockchain technology revolutionize excipient supply chain management?", *Journal of Excipients and Food Chemicals*, vol. 7, no. 3, p. 910, 2016.

[53] M. M. Queiroz, R. Telles, and S. H. Bonilla, "Blockchain and supply chain management integration: a systematic review of the literature," *Supply Chain Management: An International Journal*, vol. 25, pp. 241–254, 2019.

[54] S. E. Chang and Y. Chen, "When blockchain meets supply chain: a systematic literature review on current development and potential applications," *IEEE Access*, vol. 8, pp. 62478–62494, 2020.

[55] I. Abu-Elezz, A. Hassan, A. Nazeemudeen, M. Househ, and A. Abd-Alrazaq, "The benefits and threats of blockchain technology in healthcare: a scoping review," *International Journal of Medical Informatics*, vol. 142, pp. 104–246, 2020.

[56] S. Manti, and A. Licari, "How to obtain informed consent for research", *Breathe (Sheff)*, vol. 14, no. 2, pp. 145–152, 2018.

[57] S. R. Evans, "Clinical trial structures," *Journal of Experimental Stroke and Translational Medicine*, vol. 3, no. 1, pp. 8–18, 2010. DOI: 10.6030/1939-067x-3.1.8. PMID: 21423788; PMCID: PMC3059315

[58] M. Clarke, G. Savage, L. Maguire, and H. McAneney, "The SWAT (Study Within A Trial) programme; embedding trials to improve the methodological design and conduct of future research," *Trials*, vol. 16, no. S2, p. 206, 2015.

[59] S. Treweek, D. Altman, P. Bower, *et al.*, "Making randomised trials more efficient: report of the first meeting to discuss the Trial Forge platform," *Trials*, vol. 16, no. 1, p. 261, 2015.

[60] S. Treweek, P. Lockhart, M. Pitkethly, *et al.*, "Methods to improve recruitment to randomised controlled trials: Cochrane systematic review and meta-analysis," *BMJ*, vol. 3, no. 2, p. e002360, 2013.

[61] V. C. Brueton, J. Tierney, S. Stenning, *et al.*, "Strategies to improve retention in randomised trials," *Cochrane Database System*, 2013.

[62] A. Synnot, R. Ryan, M. Prictor, D. Fetherstonhaugh, and B. Parker, "Audio-visual presentation of information for informed consent for participation in clinical trials," *Cochrane Database System*, 2014.

[63] K. Gillies, S. C. Cotton, J. C. Brehaut, M. C. Politi, and Z. Skea, "Decision aids for people considering taking part in clinical trials," *Cochrane Database System*, 2015.

[64] R. Kumar and R. Tripathi, "Traceability of counterfeit medicine supply chain through Blockchain," In *Proceedings of the 2019 11th International Conference on Communication Systems & Networks* (COMSNETS), 2019, pp. 568–570. DOI: 10.1109/COMSNETS.2019.8711418.

[65] R. R. Konapure and S. D. Nawale, "Smart contract system architecture for pharma supply chain," In *Proceedings of the 2022 International Conference on IoT and Blockchain Technology* (ICIBT), 2022, pp. 1–5. DOI:10.1109/ICIBT52874.2022.9807744.

[66] P. Saindane, Y. Jethani, P. Mahtani, C. Rohra, and P. Lund, "Blockchain: a solution for improved traceability with reduced counterfeits in supply chain of drugs," In *Proceedings of the 2020 International Conference on Electrotechnical Complexes and Systems* (ICOECS), 2020, pp. 1–5. DOI:10.1109/ICOECS50468.2020.9278412.

[67] A. Kumar, D. Choudhary, M. S. Raju, D. K. Chaudhary, and R. K. Sagar, "Combating counterfeit drugs: a quantitative analysis on cracking down the fake drug industry by using Blockchain technology," In *Proceeding of the 2019 9th International Conference on Cloud Computing, Data Science &*

Engineering (Confluence), 2019, pp. 174–178. DOI: 10.1109/CONFLUENCE.2019.8776891.

[68] P. Sharma, M. D. Borah, and S. Namasudra, "Improving security of medical big data by using blockchain technology," *Computers & Electrical Engineering*, vol. 96, p. 107529, 2021.

[69] P. Sharma, N. R. Moparthi, S. Namasudra, V. Shanmuganathan, and C. H. Hsu, "Blockchain-based IoT architecture to secure healthcare system using identity-based encryption," *Expert Systems*, 2021. DOI: https://doi.org/10.1111/exsy.12915.

[70] S. Namasudra, P. Sharma, R. G. Crespo, and V. Shanmuganathan, "Blockchain-based medical certificate generation and verification for IoT-based healthcare systems," *IEEE Consumer Electronics Magazine*, vol. 12, no. 2, pp. 83–93, 2021. DOI: 10.1109/MCE.2021.3140048.

Chapter 4

Managing health insurance using blockchain technology

Tajkia Nuri Ananna[1], Munshi Saifuzzaman[2], Mohammad Jabed Morshed Chowdhury[3] and Md Sadek Ferdous[4]

Abstract

Health insurance plays a significant role in ensuring quality healthcare. In response to the escalating costs of the medical industry, the demand for health insurance is soaring. Additionally, those with health insurance are more likely to receive preventative care than those without health insurance. However, from granting health insurance to delivering services to insured individuals, the health insurance industry faces numerous obstacles. Fraudulent actions, false claims, a lack of transparency and data privacy, reliance on human effort and dishonesty from consumers, healthcare professionals, or even the insurer party itself, are the most common and important hurdles towards success. Given these constraints, this chapter briefly covers the most immediate concerns in the health insurance industry and provides insight into how blockchain technology integration can contribute to resolving these issues. This chapter finishes by highlighting existing limitations as well as potential future directions.

Keywords: Blockchain; Health insurance management; Customer service management; Fraud detection and risk prevention; Claim and billing management; Data record and sharing management

4.1 Introduction

Health insurance is a means of providing financial support for a person's healthcare expenses [1]. In exchange for a premium, the insured must pay all medical, surgical, and even dental treatment charges incurred by the insurer. Many organizations

[1]Department of Computer Science and Engineering, Metropolitan University, Sylhet, Bangladesh
[2]Dynamic Solution Innovators Ltd, Dhaka, Bangladesh
[3]Department of Computer Science and IT, La Trobe University, Australia
[4]Department of Computer Science and Engineering, BRAC University, Bangladesh

provide health insurance benefits to their employees, which encourages them to perform well [2]. Accidents and illnesses are unpredictable in life because they can happen to anyone at any time. The unexpected cost of treatment can put a huge strain on a person's finances. Healthcare costs are increasing rapidly as medical technology advances [3,4]. Because medical costs are going up and people want better care, more people are getting health insurance [5]. According to the study by Mendoza-Tello *et al.* [5], health insurance acts as a shield of safety in three ways. To begin with, it ensures preventative access to medicine and healthcare. When compared to uninsured individuals, those with health insurance have a better chance of accessing early detection tests for any illness. In the United States of America, lack of health insurance causes poor cancer outcomes. Uninsured individuals do not receive preventative care, which results in delayed cancer detection and a lower chance of survival [6–9]. Uninsured people are also more likely to be diagnosed at a later stage of the disease [10,11] and have a lower chance of surviving [8,11]. Second, it enables access to disease treatment that is personalized to the insured's preferences by the insurer. Insured individuals are more likely to obtain evidence-based care than uninsured individuals [9,12]. Third, it gives economic protection, discounts, and compensation for health care expenses. Insurance companies negotiate discounts with healthcare providers and thus pay a major portion of the insured's medical expenses [13]. That is how an insured person has access to reimbursement and financial stability compared to an uninsured person. An uninsured person may be forced to pay a large sum of money, resulting in a significant financial loss and health deterioration [5,14]. The health insurance issuance process starts when the customer fills out the application form in order to purchase a plan [5,15]. The insurance company that provides insurance service is addressed as the insurer. A person who purchases insurance is called an insured party. Depending on the insured's selected plan, medical background, age and insurance sum, the insurer offers the premium amount, which is paid by the insured party to the insurer [5,15]. In some situations, the insured party is required to undergo a medical examination before the insurer decides how much money to give the insured party [15]. Finally, the insurer is provided a customized policy and receives health coverage based on a mutual contract between the insurer and healthcare provider [5,15].

The health insurance industry's processes are riddled with inefficiencies, such as claim falsification, duplicate claims, fraud activities, irregular billing, security and privacy difficulties in data sharing and so on. There are numerous areas that require development in terms of efficiency, accuracy, and management. Fraud in health insurance is one of the major concerns among the issues. On September 30, 2020, healthcare insurance fraud to federal healthcare programs and private insurers have resulted in a $6 billion loss in the United States. A total of 345 individuals have charged for their roles in this fraud, with 100 of them being doctors, nurses, and medical professionals [16]. According to a survey by the National Health Care Anti-Fraud Association (NHCAA), one healthcare costs the United States more than $2.27 trillion each year, with tens of billions of dollars lost due to healthcare insurance fraud [17]. Security and privacy issues have been a major concern when it comes to health data sharing as people are extremely sensitive about their medical data, which makes customers hesitant to disclose their health data. Many people believe that the healthcare industry is not well prepared to deal with the growing number of cyber threats [18].

Various fields, such as machine learning, blockchain, and others, have had an effect on health insurance management. Machine learning is one of the most important and rapidly developing fields in the world today. Although machine learning has not been utilized extensively in the management of health insurance, there have been a few noteworthy achievements in this area. The authors of [19] have proposed an approach based on machine learning to combat health insurance fraud. Similarly, machine learning implementations have been observed in [20] and [21]. In [22], authors have leveraged data mining technologies to address fraud and claim management problems in health insurance management whereas authors of [23] have made notable contributions to the resolution of claim processing in health management. However, the most significant issue with all of these works is that privacy and security have not been adequately addressed. In addition, no mechanism for the secure exchange of data has been presented. This is a vulnerable case because privacy is the top priority for the vast majority of people today. Blockchain, on the other hand, is one of the most secure emerging technologies, and it possesses a number of characteristics that allow it to solve these significant issues. Blockchain is a distributed, decentralized, and immutable ledger technology that facilitates the transaction recording process and assists in the tracking of both tangible (cash, land) and intangible assets (intellectual property, patents, copyrights) [24,25]. It acts as a database, storing information electronically in a digital format [26]. Each transaction in a blockchain is recorded as blocks that are linked together using cryptography [27–29]. The connection between blocks provides the blockchain's immutability or resistance to manipulation by assuring that no blocks can be altered [24]. It provides immediate, shared, and completely transparent information that is stored on an immutable ledger that is accessible only to permissioned network members [24]. Blockchain technology has the characteristics of decentralization, tamper-resistance, and traceability [30]. It is capable of constructing a secure and private network as well as has the potential to resolve a wide range of health insurance issues such as interoperability, fraud activities, billing, service management and so on. Additionally, smart contracts on the blockchain can make the entire insurance process and documentation transparent. Blockchain holds all parties, including healthcare professionals, pharmaceutical companies, and insurance companies, accountable for their activities. This contributes to the development of trust among insurers and policyholders, healthcare professionals and patients as well as the overall healthcare industry [31]. As a result, it can be stated that blockchain has the capability of resolving the majority of the concerns inherent in the current health insurance system and creating a secure environment to facilitate the entire health insurance system.

This chapter reviews the most promising research work on the management of health insurance using blockchain technology. To the best of knowledge, there is not a single review paper, survey study, or even book chapter that can assist future scholars in this topic, which motivates this work. In order to classify existing works, this chapter divides health insurance management into a number of key research areas. Consequently, this would assist future researchers get a broad understanding of the field and figure out where blockchain and the existing literature fall short.

Authors' contributions: The authors' contributions in this chapter are as follows:

1. A survey on managing health insurance using blockchain technology has been conducted. This chapter assesses the techniques, their use cases, experiment scenarios, and their limitations by looking at the most relevant and recent research studies.
2. Five different aspects are then analyzed by presenting foundational demonstrations and major key concepts from existing works, and a summary was drawn.
3. This chapter concludes with a comprehensive discussion of the limitations and future directions of the reviewed papers.

The following is how this chapter is organized: it begins with Section 4.2, which discusses several key aspects and challenges of the health insurance industry. The causal effects of health insurance then are discussed in Section 4.3. Section 4.4 discusses blockchain's revolutionary impact on the health insurance industry. There is a detailed discussion of a few of the most important core factors that affect health insurance management from Section 4.5 to Section 4.9. The chapter concludes with a discussion of limitations and potential future scopes focused on the reviewed literature as well as blockchain technology in Section 4.10.

4.2 The insurance industry: its aspects and challenges

Insurance, in its simplest form, protects any individual, organization, or other entity (entity in short) against financial loss. The purpose of insurance is to safeguard any entity's financial well-being in the event of an unforeseen misfortune for instance, structural damage, loss of well-being, and so forth [32].

The insurance industry is vast, having numerous components intertwined with it. It is one of the areas where customer service management plays an important role in the overall process. If the consumer is dissatisfied with the service, the entire system suffers substantial losses. The insurance company should be able to handle the insured parties' complaints and grievances, as they play a significant role in the survival of any insurance company. Given all of these considerations, insurance should be an honest field with the goal of selflessly assisting people. But, in reality, the situation is considerably different from what was anticipated. The conviction about insurance is that if a person receives insurance and pays the monthly premium, he or she will not have to worry about paying in the event of an unexpected accident or distress. That is the insurance system's expected behavior. However, despite all of its benefits, the insurance industry has so many dark and sophisticated sides that it is difficult to comprehend. There are so many areas where the insurance sector gets compromised by being falsely used by fraud activities. The actors behind these activities can be the customer, the healthcare provider or even the insurer themselves. There are many ways the insurance sector can be compromised by an insured person or healthcare provider, which are as follows:

1. The insured party falsely claiming for additional financial assistance from the insurance agency.
2. Insured individual claiming money from the insurer for services that were never provided.
3. Falsifying facts while issuing insurance cards.
4. Healthcare providers may charge for irrelevant or unprovided services.
5. Extra money can be claimed by drugstores and healthcare providers for medicine and other items.

Insurance agencies can do many things to avoid responsibility and overcharge the customer with higher rates to tactfully gain more cash. The most common medical coverage scams that can deceive customers are listed below [33]:

1. Insurance companies may purposefully deny claims in order to create a scenario in which they are the victim and the healthcare professionals are the scammers.
2. The insurance company can change the coverage without notice in order to further their selfish desires.
3. They may charge the insured for out-of-network costs that the insured is unaware of, or they may add hidden costs to the insurance premium.

Along with these, there is widespread concern among consumers regarding the security and privacy of health data shared with insurance companies, as people are more concerned about their privacy. The method of exchanging information is inefficient and does not adequately protect individuals' privacy. Additionally, the current insurance system is so reliant on manual work that it exposes the entire system to damage. As a result, all critical aspects of health insurance require immediate attention. The core aspects of the health insurance industry are illustrated in Figure 4.1.

4.3 Causality in health insurance

In general, "casualty" refers to the link between the cause and the impact of a specific event or occurrence. Consequently, it is extremely significant in terms of health insurance. The effect of the presence or absence of health insurance on healthcare utilization and health outcomes is referred to as a casualty in health insurance. The most important elements influencing healthcare advancement are optimized access to healthcare and increased quality of care delivered [34–37]. One of the most concerning topics in the United States has been the lack of health insurance and its potential impact on healthcare utilization and health outcomes [38]. The New York Times

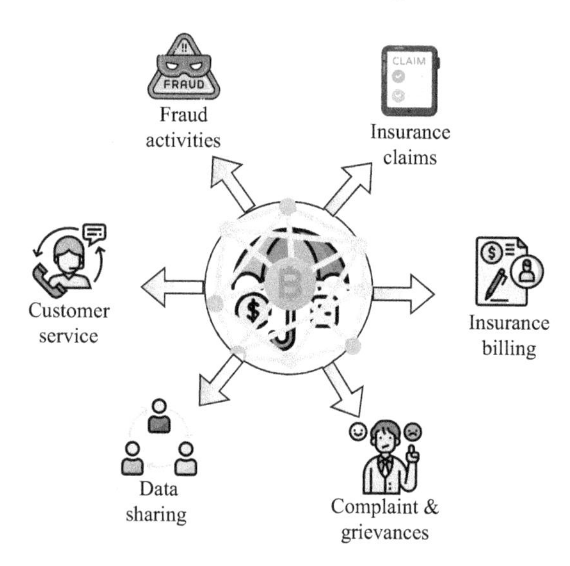

Figure 4.1 Health insurance aspects

reported a list of nearly 47 million uninsured people in the United States in 2006 [39]. In addition, the Institute of Medicine [40] and the National Coalition on Health Care [41] have recommended universal health coverage to prevent a potential health disaster.

Though most people, including citizens and politicians, believe that universal health insurance coverage is extremely important, relatively few actions have been taken to bring about this basic and significant change in practice. Few studies have addressed the relation between health insurance and healthcare access, utilization and health outcome [38].

The American college of Physicians – American society of internal medicine [42] presents the significant data it has collected in “No health insurance? It is enough to make you sick.” According to the findings, insured people have more access to healthcare, receive better hospital-based care, have a lower likelihood of having delayed treatment, and have a lower overall death rate. Brown *et al.* [43] have also summarized their work in the same way, demonstrating that insured people have better access to healthcare, resulting in a reduced mortality rate. Hadley *et al.* [44] likewise have emphasized that insured people have better access to healthcare, preventative and diagnostic services, and have lower mortality rates. Another key aspect has been discovered: insured people are more likely to be diagnosed with any illness at an early stage. Hoffman and Paradise [45] have highlighted the relationship between health insurance and health outcomes. It has been observed that insured people receive better care for chronic diseases and are less likely to encounter the obstacles of delayed treatment or missing out on necessary treatment. Howell *et al.* [46] have demonstrated the benefits of health insurance coverage for pregnant women in terms of better health outcomes before, during, and after the baby is born. Insured people also have the chance of getting regular checkups from a doctor and have a lower waiting time [49]. Table 4.1 illustrates some of the studies that have focused on analyzing the relationship between health insurance, healthcare utilization, and health outcomes.

Furthermore, it is obvious from the preceding discussion that there is a link between health insurance and healthcare utilization and/or health outcomes. When compared to uninsured people, insured people receive better treatment and facilities, as well as more preventative care. All of these amenities contribute to lower mortality rates since insured people receive preventative treatment, have better

Table 4.1 Research findings on the effects of health insurance

References	Summary[a]
American College of Physicians: no health insurance? It is enough to make you sick [42]	Insured are more likely to have proper access to healthcare and preventative services, they are less likely to receive delayed treatment, report mismanagement or missing services, receive better hospital-based care, and have a lower mortality rate
Brown *et al.* [43]	Insured have better access to healthcare and lower mortality rate
Hadley [44]	Insured receive better preventative service and access to medical care, more likely to be diagnosed at an early stage of any disease and have lower mortality rates
Hoffman and Paradise [45]	Insured have greater access to care, less likely to encounter any unmet treatment or delayed treatment, access better hospitalized care and lower mortality rate
Howell [46]	Increased health coverage for pregnant women before, during and after birth
Institute of Medicine: care without coverage [47]	Insurance provides regular healthcare facilities, an increase in preventative treatments, and proper treatment for chronic illnesses, as well as improved health outcomes
Office of Technology Assessment: does health insurance make a difference? [48]	Insured people are more likely to receive proper care and have fewer negative health outcomes
Weissman and Epstein [49]	Insured people are more likely to have a regular doctor and to obtain preventive care. They have shorter treatment wait times, are less likely to receive delayed care, and obtain better emergency care

[a]Conclusions are presented in comparison to the uninsured, with the exception of Howell (2001) and Levy and Meltzer (2001)

facilities, and can be detected at an earlier stage of any disease. As a result, health insurance improves health outcomes and healthcare utilization.

4.4 Blockchain technology: a revolution in the health insurance industry

As discussed previously, the health insurance industry is being compelled to compromise its efficiency and prosperity as a result of the high number of hazardous incidents. Blockchain technology possesses the potential to bring numerous benefits and resolve the majority of health insurance problems and concerns. The health insurance industry is being disrupted by blockchain innovation and this time for good! According to Markets and Markets [50], the global market for blockchain in insurance is predicted to grow from USD 64.50 million in 2016 to USD 1,393.8 million in 2025 [51].

So, on that basis, here are a few important benefits of blockchain in the health insurance industry:

- **Efficient data sharing:** One of the most crucial components of the blockchain is its transparency. The blockchain provides an immutable and transparent data exchange technology that assures the integrity of the data shared through the blocks. It aids insurance companies in ensuring the authenticity of shared data. Blockchain is protected by cryptographic techniques such as digital signatures and hash functions, which safeguard the process of data sharing and protect the privacy of each individual.
- **Combating frauds and false claims:** One of the most concerning hazards in the health insurance industry is false claims and fraudulent activities. The health insurance industry takes so many preventative measures in order to avoid false claims and fraudulent activities. However, the adversaries manage to deceive the insurance industry in some way. As a result of the Blockchain's ability to record time-stamped transactions with complete audit trials, counterfeiters find it incredibly hard to commit fraud [51].
- **Enhancement in customer experience:** Customer satisfaction is a critical component of any insurance company's sustainability. Customers are typically content with the provider for whom they pay lower premiums and receive enhanced benefits. Additionally, if an insurance agency is capable of responding quickly to customer concerns and providing justice to any client, this contributes significantly to strengthening consumer trust in that insurance company. However, maintaining all of these simultaneously is extremely difficult. One solution is to leverage blockchain technology to automate processing through the use of smart contracts. Business agreements are encoded in the blockchain, and payments are triggered and processed instantly [51].
- **Improves trust among entities:** In this scenario, the blockchain smart contract emerges as a savior by establishing trust throughout the formation of the insurance contract. In every transaction, smart contracts provide immutability

and auditability. In addition, it eliminates the need for an intermediary to manage the insurance process.

- **Collect and store data:** The insurance industry is data-driven. Using technologies such as artificial intelligence (AI) and the Internet of Things (IoT), blockchain technology enables the collection of a wide variety of valuable data. IoT-collected data is stored on the blockchain and then analyzed by AI. This enables the company to take an informed and autonomous decision on insurance premiums [51,52].
- **Accountability and ownership:** Accountability allows all parties involved in a commitment to be open and honest with one another. The immutability of block transactions and their connections enables ownership control and accountability [53]. Using the open and decentralized nature of blockchain, customers can see what the insurance company knows about them and how that information is used, which means that there is more openness and honesty [54].

According to Kuo *et al.* [55], the major benefit of blockchain integration for the healthcare industry is the decentralization behavior of blockchain, which allows relevant parties such as healthcare providers, insurance companies, and government regulatory bodies to share data securely without relying on any central system. Immutability, transparency, security, and robustness are other important factors in making any blockchain-based system more efficient and improving data sharing. With such exceptional benefits, it may be concluded that blockchain is the optimal way to help the health insurance industry. In later sections, the most crucial aspects of health insurance are investigated thoroughly, along with how blockchain can provide a substantial path to resolving these issues.

4.5 Customer service management of health insurance using blockchain technology

Customer service is critical in developing a positive relationship between a company and its customers. To be specific, a company's success is heavily reliant on customer service and satisfaction. Customer service is the assistance provided by enterprises before, during, and after the purchase or use of their products or services. Medical aid supplied to patients, any type of insurance service provided by insurance firms, or any materialistic product sold by companies are all instances of customer service. Good customer service improves customer satisfaction and enterprise performance, which leads to the enterprise's ultimate success [56].

The quality of customer service is not just determined by the customer service. It is heavily dependent on how the organization uses its current customer service scenario to attract new customers and retain the loyalty of existing ones [57]. Customer service is a kind of knowledge-based work where the main motive is to efficiently utilize and accumulate previous knowledge to improve the customer service quality [58]. Therefore, it can be stated that an efficient customer service management system is an essential component of every company's success.

Health insurance provides customer service to its members who purchase health insurance from their company. The objective of a health insurance company is to provide financial benefits to its customers. An insured person receives health coverage as a result of the insurer's and health-care providers' combined efforts. However, managing customers and providing adequate customer care is a difficult task. The entire system is entirely dependent on human supervision, making it prone to error. Other drawbacks include the insured individual fabricating information, abnormal billing, a lack of transparency, and, most crucially, a lack of data privacy and security. These difficulties must be addressed in order for the health insurance organization to be successful. Blockchain has the ability to alleviate the majority of the problems associated with the health insurance customer administration system. Blockchain technology can provide a secure environment in which the health insurance management system can be easily integrated.

According to the preceding discussion, customer service management is about providing quality service to existing customers as well as attracting new customers with better service. Considering this information, it can be stated that there are some factors that influence quality customer management, such as thoroughly monitoring if current customers are receiving proper facilities; how the insurance company responds in the event of any obstacles encountered by the customer; and, most importantly, if any customer's complaint is being resolved patiently. As a result, customer management in health insurance falls into two broad categories: *monitor and response management* and *complaint and grievance Management*. These two aspects are among the core aspects of health insurance management, as demonstrated in the following sub sections.

4.5.1 Monitor and response management

Monitor and response management in health insurance play a vital role in ensuring quality customer service. Monitor and response management in health insurance refers to the ability to properly manage health insurance through monitoring and ensure better quality service by providing a proper response to customer demands and issues.

The insurance industry's major step in this regard is to constantly monitor all of their customer facilities, such as whether customers are receiving all of the services included in their insurance policy. If this factor is not ensured, achieving existing customer satisfaction can be extremely difficult, let alone attracting new customers. However, this aspect is more than just constantly monitoring existing client facilities; it serves a larger purpose. With medical fields getting better and new cures and devices being made almost every day, the cost of medical care and getting access to it has become hard to handle. As a result, insurance companies must monitor their policies on a regular basis and, if necessary, reform them. Otherwise, maintaining the standard and remaining among the best companies is extremely difficult.

The main challenge in ensuring these aforementioned factors is ensuring quality service by providing adequate privacy to customers. As monitoring and response management are essentially about sharing information about customers, it is critical to protect each customer's privacy. Customers must also be assured that their privacy is not violated. Current systems, however, struggle to meet these requirements due to inconsistencies in security policies and access control structures. Blockchain paves the way for a revolution in this field. If the privacy and transparency properties of blockchain can be efficiently integrated into solving these issues, it could be a massive benefit and make the monitoring and response management process run smoothly.

4.5.2 Complaint and grievances management

It is crucial how a health insurer handles customer complaints and concerns. The quality of their service is highly dependent on how they handle their customers' concerns [59]. Customer complaint resolution facilitates the resolution of unsatisfactory situations and the tracking of complaints and grievances, paving the way for an organization to enhance its service quality [60,61].

Any individual has the freedom to select any insurer he or she wishes, based on the insurer's service quality. Customers have the option of switching plans or even insurance companies if they are displeased with the outcome of their complaints [62]. In typical health insurance complaint and grievance management systems, the client must physically attend in order to file a complaint or even monitor its status. Even if the customer's complaint is addressed, he or she might be dissatisfied with the entire system if the settlement is too lengthy. Additionally, there is the concern of transparency and anonymity. In today's society, people are extremely protective of their data, which makes them feel uncomfortable filing a complaint just because the system lacks anonymity. There may be additional causes for ineffective complaint and grievance management, such as a failure to enforce robust complaint and grievance management laws and regulations or a flawed redressal system [63].

To the best of our knowledge, there is no literature that directly addresses health insurance complaints and grievances. As a result, similar fields have been investigated how research communities have addressed complaints and grievances. Health insurance companies can utilize similar solutions to manage their clients' complaints and grievances, ensuring that their company's quality continues to improve.

There is a lack of communication between the government and its citizens. In some parts of the world, the manual process of filing a grievance can take up to a month. Furthermore, tracking the status of a complaint is a time-consuming and tiresome task. With these problems in hand, Jattan *et al.* [64] have used the ethereum blockchain to propose a secure and transparent system for resolving complaints. Each complaint is a smart contract processed on the ethereum Blockchain. The process begins with a user registering a complaint. The complaint is registered and so kept on the ethereum blockchain by including the appropriate details. To track the status of the complaint, the user will be granted a complaint number.

Table 4.2 Key findings in complaint and grievances management

Reference	Key finding	Advantages	Disadvantages
Jattan *et al.* [64]	Proposed a secured and transparent BC system for complaint redressal system using Ethereum.	• Secure • Transparent and immutable	• Applicability • Scalability • Benchmarking
Rahman *et al.* [65]	Presented a BC-based anonymous and transparent platform where complainants can submit anonymous complaints and communicate with authorities for resolving their complaints.	• Secure and reliable • Ensures anonymity of users • Efficient • Ensures temper resistance	• Simulation • Privacy • Efficient • Benchmarking

Officials are able to view the complaint only if the data is stored on the blockchain and take appropriate action on it. The complainant will be kept informed of all actions taken in response to the complaint.

Quick and effective complaint resolution is a critical civic right that every citizen expects from their government. In a typical physical complaint management system, people must physically visit the organization to file a complaint. These complaints are addressed in a committee-based system, which is a really long and tiresome process. There are various online complaint management tools accessible. Most victims are hesitant to submit a complaint through these online platforms due to the lack of transparency.

In [65], Rahman *et al.* have presented a blockchain-based platform for creating an anonymous, transparent, and decentralized environment for complaint and grievance management. Individuals can use this method to lodge anonymous complaints and communicate with the authorities responsible for resolving their problems. This platform enables officials to deal with complaints more efficiently. It makes use of self-sovereign identity and zero-knowledge proof to ensure that users' identities remain anonymous while lodging complaints, thereby reducing their vulnerability to threats. As a storage and sharing mechanism, this platform makes use of the interplanetary file system, which, like blockchain, is resistant to tampering and provides additional security.

Summary: To assist future researchers, the essential concepts of existing works have been outlined in Tables 4.2 and 4.3. In contrast to the work of Jattan *et al.* [64], the work of Rahman *et al.* [65] is reliable, efficient, and ensures user anonymity. Despite the fact that both existing works are built on public blockchain, Jattan *et al.* have used ethereum, whilst Rahman *et al.* have used hyperledger aries and hyperledger indy.

Table 4.3 Different utilized properties in complaint and grievances management (– indicates that the required information is missing/not mentioned)

Reference	Blockchain type	Implementation language & other platform/framework	Consensus algorithm	Blockchain platform
Jattan *et al.* [64]	Public	Next.Js, Web3.js	–	Ethereum
Rahman *et al.* [65]	Public	–	–	Hyperledger aries and hyperledger indy

4.6 Health insurance claim management using blockchain technology

In the context of health insurance, a claim is any application filed for benefits from the health insurer organization. The client must file a claim so that funds for his or her care can be reimbursed. The health insurance provider must process the claim request in order to manage claims. This means that the insurer must investigate the claim request's authenticity, rationale, and information. There are various stages involved in the processing and management of claims. This process begins with registering the customer, continues with policy issuance, ensures the customer's authenticity by verifying medical certificates [66], keeps customer data confidential, detects false and anomalous claims, and reduces overall management costs [67]. If everything is in order, the health insurance provider reimburses the client's health care provider.

However, the major challenges in claims management include time management, reliance on human supervision, a lack of visibility, and security. Because most management systems rely on human supervision, the claim management and insurance issuance processes are prone to errors. The data collection and processing procedure is entirely dependent on humans, making it prone to inaccuracy [5]. Also, a major constraint with this entire process is the lack of data visibility and availability and security. Consequently, obtaining health insurance and managing claims becomes a laborious and time-consuming process for both the insurer and the insurance provider [5,67]. Another significant issue in insurance claims is that the insurance process is occasionally hampered by service manipulation and misuse by policyholders and providers seeking a higher payment for an insured incident [5]. Because of these concerns, it has become difficult to ensure the integrity and legitimacy of the claim request information [5,67].

With this concern growing in health insurance claims, there is various research work that has focused on building tempering-free claim architectures. Thenmozhi *et al.* [67] have proposed their architecture based on a major objective: develop a blockchain-based health insurance claim processing system that will assist insurance companies in establishing a secure, tamper-resistant network. The proposed system consists of three modules, including:

1. **Registration:** When a person joins the organization by paying the annual premium, the system calculates the required premium. A hospital joins every year and its stay or drop depends upon the voting system. It can enter its offers into the system. Patients can browse offers and choose treatments.
2. **Treatment:** A person chooses a hospital and an offer from the organization's interface (web). The hospital records the person's treatment in the system. The system stores the record.
3. **Claim:** Payment is made in the hospital following treatment. Due to the fact that the record is still open and not closed, the organization calculates the payable amounts. The patient can access his or her personal information, payment history, and payable balance via the user web portal.

Authors from [5] have highlighted the characteristics of blockchain technology that are used for issuing health insurance contracts and claims, including automation, authentication [68], transparency, immutability and decentralization. A layer model is defined based on these characteristics to abstract the functionality of the schema components, namely peer-to-peer mining, blockchain, smart contract and user. As a result, the studies have noted three use cases: *the issuance of an insurance contract (policy), payment of an insurance premium, and claim management.*

Transparency at all levels is absolutely critical to health insurance and more specifically, the health sector. In [31], Sawalka *et al.* have proposed a blockchain-based claim model to ensure transparency between insurance companies, obviating the need for agents and enabling direct contact between insurance companies and hospitals. EthInsurance, the proposed framework, enables efficient and secure access to medical data by patients, providers and other third parties while safeguarding patient privacy.

EthInsurance consists of three modules i.e. *the patient, the hospital and the insurance company*. The blockchain is intrinsically linked to the patient and it determines who is granted access. Three contracts, namely *consensus, permission, and service*, are in charge of all blockchain activity. Experiment results show that EthInsurance is reliable and safe. Utilization of blockchain technology lowers the cost of decentralization. It validates and authorizes data before it is transmitted over the network, minimizing the possibility of unauthorized use of records. Furthermore, each patient has a unique ethureum address and identifier. The authors have used distinct contracts to convey a sense of modularity, which benefits the framework's data security.

Summary: An overview of the key contributions, benefits and numerous utilized components of the most significant works is provided in Tables 4.4 and 4.5. All existing mechanisms are secure, reliable, and reliable. By facilitating direct contracts between entities, Sawlka *et al.* [31] have eliminated the necessity for agents. Mendoza-Tello *et al.* [5] have, on the contrary, increased transaction durability. Authors from [67] have reduced network latency in order to construct a safe, tamper-free network.

Table 4.4 Key findings in claim management

References	Key finding	Advantages	Disadvantages
Sawalka *et al.* [31]	Proposed a BC-based system that stream-lines and simplifies the insurance claim process	• Eliminates the need for agents • Enables direct contract between entities • Secure and efficient access to medical data • Reliable and safe	• Enhancement Benchmarking
Thenmozhi *et al.* [67]	Developed a BC-based solution to aid insurance firms in establishing a tempering-free secure network	• Secure and reliable • Network latency	• Applicability • Scalability • Security • Benchmarking
Mendoza-Tello *et al.* [5]	A layer-based model is built and the three usage scenarios are presented based on the characteristics provided by BC	• Transaction durability • Trust • Reliable and safe	• Feasibility • Benchmarking

Table 4.5 Different utilized properties in claim management (– indicates that the required information is missing/not mentioned)

References	Blockchain type	Implementation language & other platform/framework	Consensus algorithm	Blockchain platform
Sawalka *et al.* [31]	Public	Ganache platform	–	Ethereum
Thenmozhi *et al.* [67]	–	Express	Proof of work	Ethereum, hyperledger fabric
Mendoza-Tello *et al.* [5]	–	–	–	–

4.7 Health insurance bill management using blockchain technology

Billing is the procedure by which a third party, typically insurance companies, reimburses a patient for treatment delivered by any healthcare provider. This entire process is referred to as the *billing cycle* or *revenue cycle management*, and it incorporates the management of claims, payment, and billing [69].

Claims and billing are core aspects of the health insurance system and are closely intertwined. The whole process starts with the doctor visiting the patient and assigning diagnosis and procedure codes, which are then used by the insurer in order to define the coverage and medical necessity of the services [70]. The patient files a claim with the insurance company, and the insurance company verifies the correlation between the services provided by the healthcare provider and the percentage they must pay. Depending on the outcome, the insurance pays the amount to the healthcare provider on the patient's behalf [71]. The insurer must verify if there is any relevance between the service provided by the healthcare provider, the necessity of those medical treatments and the amount of finances demanded in the claim. The reason behind this procedure is that there can be fraud cases which can be executed by the healthcare provider for higher benefits or even the patient themselves. Customers can deceive insurers by claiming for services that were never delivered, for unneeded medications, for repeated claims, or even for fraudulent insurance cards. Similarly, healthcare professionals may charge a fee for any service or treatment that was not provided, or for providing medically excessive and unneeded services [72]. As a result, billing management is a crucial part of the insurance system for the seamless operation of the whole process. For a long time, health insurance companies managed their whole billing system using pen and paper. Currently, the majority of companies manage claims and billing through the use of digital technologies.

To the best of our knowledge, billing management in health insurance has not been addressed separately over the years. Instead of that, research communities focused on this aspect during fraud and claim management activities. As a result, additional domains have been searched in which researchers have addressed billing management in order to reduce fraudulent activity and streamline the claim process.

Authors from [73] have proposed a blockchain-based approach that uses Internet of Things (IoT) devices for the metering and billing of customer for the electric network. The proposed mechanism aims to solve both rust and privacy issues. They used raspberry pi to simulate metering and hyperledger fabric, with its decentralized structure, to provide transparent and safer solutions. Experimental results show the proposed system is less vulnerable to cyber attacks as a consequence of using blockchain. In addition, by implementing smart contracts and automating all energy tracking procedures, the rate of wrong manual measurements will be reduced.

Sawalka *et al.* [31] have aimed to make the insurance process more seamless and less time-consuming. They use insurance as a mode of payment at the hospital. The hospital directly requests a claim from the company, and the insurance company responds with a claim record after checking details in the patient's records. This system reduces the patient's workload while also keeping the process transparent. This paper's details have been skipped because they have already been discussed in Section 4.6.

Summary: Tables 4.6 and 4.7 provide a summary of the findings of the reviewed papers, which might help researchers gain a rapid understanding of what previous researchers have accomplished. Both the mechanisms from Ahmed *et al.* [73] and Sawalka *et al.* [31] have given secure and efficient data access, hence

Table 4.6 Key findings in billing management

References	Key finding	Advantages	Disadvantages
Ahmet *et al.* [73]	A consumer metering and billing system is presented that uses BC technology and IoT devices to address scalability, trust, and privacy concerns	• Scalable energy efficient • Less vulnerability to cyber • attacks • Privacy	• Security • Privacy
Sawalka *et al.* [31]	Proposed a BC-based system that streamlines and simplifies the insurance claim process	• Eliminates the need for agents • Enables direct contract between entities • Secure and efficient access to medical data, reliable and safe	• Enhancement • Benchmarking

Table 4.7 Different utilized properties in billing management (– indicates that the required information is missing/not mentioned)

References	Blockchain type	Implementation language & other platform/framework	Consensus algorithm	Blockchain platform
Ahmet *et al.* [73]	Private	Javascript	Proof of concept	Hyperledger fabric
Sawalka *et al.* [31]	Public	Ganache platform	–	Ethereum

reducing the exposure to cyber threats. In addition to the comparison, Ahmed *et al.* have employed *hyperledger fabric*, while Sawalka *et al.* have employed *ethereum.* As part of the implementation, Ahmed *et al.* utilized *javascript* for their private blockchain, whereas Sawalka *et al.* leveraged the *ganache platform* for their public blockchain.

4.8 Fraud detection and risk prevention in health insurance using blockchain technology

Fraud is the intentional deception or manipulation of information by an individual or organization in order to obtain a financial or personal advantage [74]. Healthcare is a prime target for adversaries to attack and profit from. Among these attacks, healthcare insurance fraud has been a significant source of burden on the healthcare industry in recent years. It has attracted the attention of the government and health insurance companies due to the significant losses it causes them [75]. Faking information, hiding third-party liability, and falsifying electronic bills are all examples of common health insurance fraud. These types of instances can be incredibly harmful to an authentic customer, both physically and financially [30].

The traditional health-care system is built on trust. The patient who provides the insurance card is trusted by the health insurance provider. The service providers, understandably, believe that the patient did not falsify the insurance card. In the event of doubt, they have to go to the insurance company physically. Furthermore, employees cannot be trusted because fraud can occur from within the insurance organization. As a result, the old system cannot be trusted as a system, which leads us back to the issue of trust [75]. The issues of privacy and security are inextricably linked to this trust issue. People are always concerned about their security and privacy before they are concerned about anything else. They will never trust any organization if their security is not ensured. As a result, people's trust in the healthcare system is deteriorating. Also, it is very frustrating considering the time, effort and money wasted on the paperwork of the whole manual process [75]. Calculations indicate that American insurance firms waste up to $375 billion per year as a result of the paperwork and administrative box-ticking [76].

Another significant issue with the current health insurance sector is that the claims are submitted manually to the insurance providers. As a result, if there are potential fraud cases, they often go unnoticed. Furthermore, the manual process often includes private individuals whose responsibilities include detecting fraud and abusive instances and proving them to the government or other regulatory authority. If they achieve their goal, they are financially rewarded, which motivates them to perform their duties more sincerely [77]. Still, the manual system is prone to errors. Many fraud instances go undetected owing to a lack of appropriate proof, resulting in massive economic loss [74]. Additionally, the current manpower and resources available for healthcare insurance are insufficient to prevent these fraudulent instances [30]. Furthermore, policyholders and medical service providers sometimes take advantage of a customer's health insurance benefits through falsification and service abuse to obtain additional funds [19]. In Figure 4.2, different aspects of fraudulent incidents in healthcare insurance have been illustrated.

Taking various forms of healthcare insurance fraud in mind, Liu *et al.* [30] have proposed a blockchain-based anti-fraud system for healthcare insurance that consists of *a cloud platform, a network layer, a core layer, an interface layer, and an application layer*. The main objective of this system is to verify the patient's medical reimbursement request and ensure it complies with the policy's provisions. The proposed architecture includes three blockchain services to address three types of fraud, namely *medical process information inspection*, *third-party responsibility inspection*, and *healthcare insurance bill inspection*. From the experiment results, the authors have shown that the proposed system is tailored to business needs and is based on resolving current real-world problems. It does not require any modifications to the existing information system. Additionally, the architecture is used to call services between systems, decouple systems, and internal modules are also a loosely coupled architecture of building blocks that can adapt to changing business

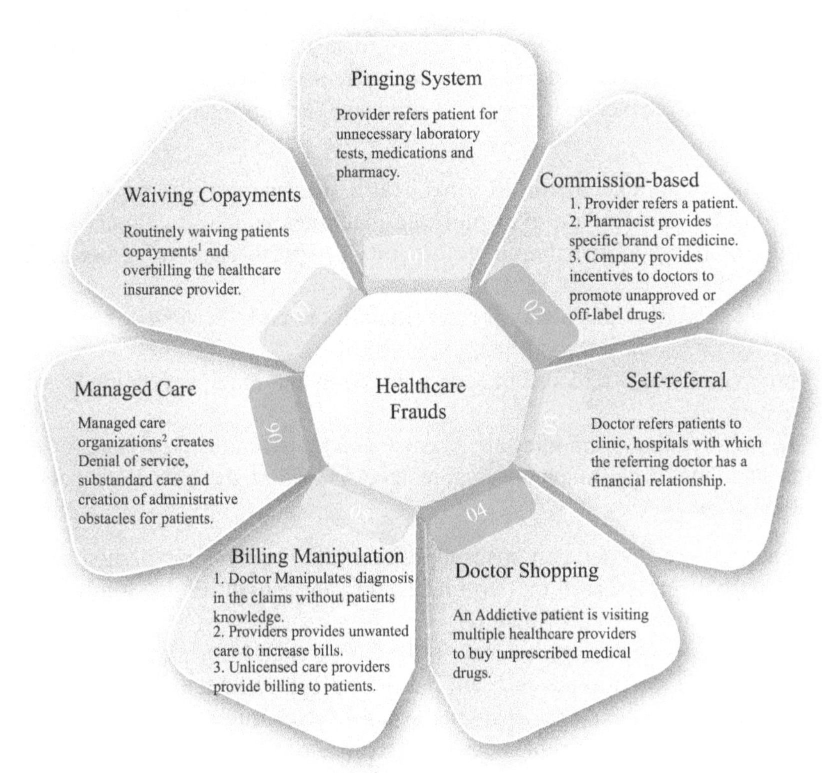

Figure 4.2 Different aspects of health insurance fraud (inspired by [74]). [1]Patients pay fixed amount to providers as defined by healthcare insurance policy. [2]An entity that connects the healthcare insurance providers and the insurers (patients).

requirements. The capability of solving anti-fraud issues, such as fraudulent data, concealing third-party liability accident fraud, false electronic bill reimbursement and other issues has made the architecture suitable for the current trend of medical information.

In [75], Alhasan *et al.* have highlighted the trust between patients and service providers, which leads to counterfeit (fraud) in the health insurance industry. Governments are spending countless amounts of money and time to stop this dilemma. To prevent this, authors have discussed recent articles that looked into various health insurance systems based on blockchain technology and hence proposed a new framework and consensus algorithm to ensure the security and decentralization of a distributed ledger. The proposed BC-based framework is composed of five major components: the insurance company, the patient, treatment providers (hospitals, medical centers, and pharmacies), the ministry of health (MOH), and the blockchain network.

The proposed system is evaluated in terms of validation time, upload time, time required to append blocks to the chain, data integrity verification, and data privacy. Experiment results demonstrate the system's robustness and effectiveness in terms of performance, security, and privacy. The advantages are highlighted below.

- The use of blockchain in storing health information can be effectively secured by having data over multiple machines which are supervised and authorized by a distributed community in preference to a centralized approach.
- This method provides a way for everyone in the party to view and verify the data that is added and modified.
- Moreover, there is a record of each transaction and modification made within the network.
- The performance of middleware to parse and transform medical health data is fast and there is not much observable delay to loading that processed data into blockchain.

Additionally, the proposed consensus algorithm reinforces the distributed concept by selecting randomly based on two rules: FIFO and length results, implying that the system is completely distributed.

Each year, the healthcare industry in the United States loses tens of billions of dollars to fraud. Certain types of fraud put the patient's health at risk. This drives Ismail *et al.* [74] to develop a system capable of detecting and preventing fraud based on twelve different fraud scenarios. Because of the peer-to-peer distributed nature of blockchain, they have proposed *Block-Hi*, a blockchain-based health insurance fraud-detection system. Furthermore, they have created a taxonomy of health insurance fraud based on various fraud scenarios as well as relationships between insurance claim contents, associated fraud categories, and corresponding validators.

The authors have investigated the performance of their system in terms of execution time and data transfer amount when the number of claims and the number of healthcare insurance company branches increased. When the number of health insurance branches increases, the performance of *Block-HI* in terms of execution time and data transferred degrades by only 0.69% on average. While the number of claims have increased, performance declined by 33.51%. This is due to the consensus protocol's execution.

Summary: To conclude this section, Tables 4.8 and 4.9 highlight the findings on fraud management using blockchain technology. In addition to providing a blockchain-based solution, Ismail *et al.* [74] have presented a potential taxonomy of fraud attacks in the healthcare insurance industry. The authors have included PBFT into their private blockchain. On the other hand, Liu *et al.* [30] facilitate enhanced system performance. The architecture of Alhasan *et al.* [75] has improved transaction speed and usability. They have introduced a new consensus method for their consortium blockchain.

Table 4.8 Key findings in fraud detection and risk prevention

References	Key finding	Advantages	Disadvantages
Liu *et al.* [30]	Proposed a healthcare insurance anti-fraud system based on BC and cloud computing that reduces the need of resources and manpower and provides various medical services	• Secured • Protects privacy of three-chain data • Supports better system performance	• Enhancement • Benchmarking
Alhasan *et al.* [75]	Proposed and implemented a novel framework and consensus algorithm to eliminate fraud in the health insurance industry	• Usability and efficiency • Effective in terms of security • Privacy and speed	• Applicability • Scalability • Benchmarking
Ismail *et al.* [74]	Presented a taxonomy of healthcare insurance claim frauds and proposed and evaluated a BC-based healthcare insurance claims fraud detection framework	• Performance enhancement in terms of execution time and the amount of data transferred	• Simulation • Feasibility

Table 4.9 Different utilized properties in fraud detection and risk prevention (– indicates that the required information is missing/not mentioned)

References	Blockchain type	Implementation language & other platform/framework	Consensus algorithm	Blockchain platform
Liu *et al.* [30]	Private	–	–	Hyperledger fabric
Alhasan *et al.* [75]	Consortium	C#	Self proposed	–
Ismail *et al.* [74]	Private	–	PBFT	–

4.9 Health insurance data record and sharing management using blockchain technology

Patient health data is one of the most powerful weapons in the health insurance sector. As a result of technology breakthroughs such as the invention of the IoT and wearable devices, healthcare data is exploding [78]. This data can provide useful insights to the health insurance company, which can subsequently use it for a variety of purposes, including policy customization and claim management [79,80]. Health insurance companies can use IoT devices to collect data in real-time and build precise and customized policies based on individual lifestyles [80,81]. As a result, it might be said that a secure data sharing infrastructure is required for sharing health data with a healthcare insurance provider in order for the consumer to obtain the full benefits of health insurance.

However, there are several obstacles, such as data privacy, security, and interoperability, that can have a significant impact on the entire data sharing architecture [81,82]. To begin with, there is concern about client data privacy and security, as health data is extremely sensitive by nature and contains personally identifiable information [83]. If this data is compromised, serious financial and physical consequences may occur. As a result, the current data sharing architectures pose a reliability concern [81,82]. Second, effective data integration and interoperability amongst healthcare systems remain a major challenge [81,82]. Another challenge is that customers have little control over their private health data [84]. All of these obstacles necessitate the development of a secure and private infrastructure for data sharing so that data can be shared with proper access control and privacy.

To address the inefficiency and time-consuming nature of the current system, Lokhande *et al.* [79] have devised a new method of sharing health data that makes use of permissioned blockchain technology to safeguard the data and conceal it from those who do not wish to see it. Additionally, authors have utilized the participation service, which is provided by blockchain to assist with distinctiveness management. Their designed mobile healthcare system can collect, distribute and collaborate on individual health data between entities and insurance companies.

The authors of [80] have proposed a method for sharing personal data while maintaining privacy and security through the use of blockchain technology. The proposed novel framework integrates health insurers, IoT-based networks, and blockchain technology to implement an access control protocol based on a smart contract for sharing insureds' financial premiums with stakeholders acting as non-participants or authorized parties. In comparison to traditional data-sharing systems, the proposal evaluation results in authorized access in less time. The proposed method, according to the security analysis, ensures transaction integrity via SHA-256, user authenticity via asymmetric encryption, non-repudiation, avoids single point failure, and privacy.

As the volume of healthcare data increases, issues with unstandardized data formats and provenance become more vulnerable. Lee *et al.* [82] have proposed a standards-based sharing framework, SHAREChain, which incorporates two features to deal with reliability and interoperability issues. These are:

1. It enhances reliability by leveraging the data integrity of a blockchain registry and establishing a consortium blockchain network to exchange data between authenticated institutions.
2. The second feature enhances interoperability with standards relating to healthcare data sharing. To ensure data interoperability, the authors have used FHIR as well as XDS's actor and transaction concepts in the system architecture.

Summary: Tables 4.10 and 4.11 provide a comprehensive summary of the overall elaboration of this aspect. Compared to prior publications, [79,80] authors

Table 4.10 Key findings in data record and sharing management (– indicates that the required information is not found)

References	Key finding	Advantages	Disadvantages
Lokhande *et al.* [79]	Proposed a system which is capable of exchanging health data by using permissioned blockchain to ensure security and concealment	• Highly ascendable • Trusted and liable	• Benchmarking
Iman *et al.* [80]	Presented a novel framework that uses a smart contract to implement an access control system for sharing the financial premiums of insureds with stakeholder as non-participants/ authorized parties	• Protect data from potential threats • Authorized access within less time • Off-chain db	–
Lee *et al.* [82]	Proposed a sharing framework which deals with reliability and interoperability issues by utilizing the data integrity and consortium BC network.	• Interoperability • Identity • On/Off-chain	• Emphasize on key factors

Table 4.11 Different utilized properties in data record and sharing management (– indicates that the required information is missing/not mentioned)

References	Blockchain type	Implementation language & other platform/ framework	Consensus algorithm	Blockchain platform
Lokhande *et al.* [79]	Private	–	Top score, proof of veracity and authentication	Hyperledger fabric
Iman *et al.* [80]	Private	HTML, CSS, JS, ASP.net, C#	PBFT	–
Lee *et al.* [82]	Consortium	XDS, FHIR	–	–

have provided trustworthy and reliable techniques. The mechanism from Iman *et al.* [80] only supports off-chain databases, but the mechanism from Lee *et al.* [82] supports both on-chain and off-chain databases. Iman *et al.*'s technique requires less time due to their usage of PBFT as their private blockchain's consensus algorithm. Lokhande *et al.* have worked with private blockchains as well, but their consensus techniques are top score, proof of veracity, and authentication. SHAREChain, the consortium blockchain technology proposed by Lee *et al.*, is implemented using *XDS* and *FHIR*.

4.10 Limitations and future directions

Blockchain technology has the potential to revolutionize the health insurance sector by resolving complex issues such as claim management and fraud detection. Researchers are recognizing the benefits of integrating blockchain in resolving these challenges of the health insurance sector. Numerous notable research projects have already been suggested and have been demonstrated to be successful in fixing the challenges for which they were established. However, many systems have one or more drawbacks that must be addressed, and additional research must be conducted to solve these limitations.

4.10.1 Limitations

In this subsection, an attempt has been made to highlight the key drawbacks of the reviewed literature in this area. A demonstration of the limitations of the existing literature is given in Table 4.12.

- **Lack of literature:** Blockchain technology is still in its nascent stage. The adoption of this technology is low in comparison to the number of problems it can solve. Despite the fact that the health insurance industry has integrated blockchain in numerous areas, there are still significant areas for which no blockchain-specific solutions exist. Health insurance billing management and complaint and grievance management are two sectors where high-quality blockchain-based solutions are needed. While billing is slightly related to the concept of claiming, it has a significant impact on its own. Although several writers have highlighted the billing process in connection with the claim management process, relatively little study has been conducted on the blockchain-based health insurance billing system. In addition, there is no research on the subject of monitoring and response management, which is highly alarming and should be addressed immediately.
- **Lack of applicability and scalability:** As previously noted, researchers can employ blockchain-based solutions that have already been implemented in other industries to address health insurance challenges. Nonetheless, this poses two crucial questions:

- Are these approaches sufficient for addressing health insurance concerns?
- Are these methods capable of handling health data, which contains sensitive and personally identifying information?

For instance, the authors of [64] have developed a method to control complaints and grievances between the government and the people. They have incorporated blockchain, and the use of blockchain has increased the complaint management system's security, transparency, and immutability. However, they have not provided evidence of the scalability or adaptability of their system, making it extremely difficult to implement the recommended strategy in the health insurance industry or any other sector. In addition, the authors from [67] have failed to clarify the adaptability and performance comparison of their systems, which are

Table 4.12 Limitations founded in reviewed literature (● means limitation exists and ○ means it does not)

Limitations in references	Applicability	Scalability	Security	Simulation	Privacy	Feasibility	Enhancement	Emphasize on key factors	Benchmarking
Sawalka *et al.* [31]	○	○	○	○	○	○	●	○	●
Thenmozhi *et al.* [67]	●	●	●	○	○	○	○	○	●
Mendoza-Tello [5]	○	○	○	○	○	●	○	○	●
Mangaonkar [85]	○	○	○	○	○	○	○	○	●
Goyal *et al.* [86]	○	○	○	○	○	○	○	○	●
Jattan *et al.* [64]	●	●	○	○	○	○	○	○	●
Rahman *et al.* [65]	○	○	○	●	●	○	○	○	●
Ahmet *et al.* [73]	○	○	●	○	●	○	○	○	○
Liu *et al.* [30]	○	○	○	○	○	○	●	○	●
Alhasan *et al.* [75]	●	●	○	○	○	○	○	○	●
Ismail *et al.* [74]	○	○	○	○	○	●	○	○	○
Lokhande *et al.* [79]	○	○	○	○	○	○	○	○	●
Iman *et al.* [80]	○	○	○	○	○	○	○	○	○
Lee *et al.* [82]	○	○	○	○	○	○	○	●	○

critical needs for any insurance firm seeking to use their proposed architecture in a real-world scenario. Alhasan *et al.* [75] have proposed an excellent method for resolving health insurance fraud events, but they have not described the scalability of their method.

Before even considering integrating an existing solution from another industry into the health insurance system, the above two concerns must be answered. Consequently, there is insufficient study to fully comprehend the applicability and scalability of existing solutions in the health insurance industry.

- **Lack of simulation and benchmarking:** Simulation and benchmarking play a crucial part in the success of any scientific study. Most research papers that use blockchain technology to solve problems with health insurance do not have quality benchmarking.

  The authors of [65] offer a significant method for managing complaints and grievances using blockchain technology. However, the authors have not provided a simulation of their proposed system, which is necessary if this strategy is to be successfully integrated into the health insurance system. Similarly, Alhasan *et al.* [75] have offered a practical and effective strategy for addressing the fraudulent concerns that occur in the health insurance market, but they have not presented any benchmarks to evaluate their work. The same restrictions apply to Ismail *et al.* [74], who have not compared the energy consumption and performance of different consensus procedures in Block-HI.
- **Lack of feasibility:** Researchers have presented a variety of potential solutions for health insurance-related problems. However, several of them are missing the feasibility study component. The purpose of a feasibility study is to determine whether the proposed solution to a problem is sufficient. For instance, the approach presented in [5] can be quite effective in the processing of health insurance claims. Though the benefits of incorporating blockchain into claim management are evident, rules cannot be established for every component of any system, especially for the extremely uncertain fraudulent scenarios in claim administration. In some instances, a sophisticated medical treatment must be rigorously validated, which demands a significant deal of experience and expertise. Similarly, Ismail *et al.* [74] have presented a solution to the problem of health insurance fraud. To achieve this, they have divided fraudulent actions into 12 distinct groups. However, manually identifying activities is quite difficult due to the possibility of unanticipated challenges.
- **Security:** System security is a significant consideration while attempting to resolve health insurance problems. A secure environment is one of the grounds for implementing blockchain in health insurance. However, while many of the ideas presented by researchers have the ability to tackle the problems, they also pose significant security risks. The solution given by [67] for the claim management aspect creates serious security problems. The problem with this study is that the database used in the suggested architecture has not been encrypted. The most promising billing management study discovered is a blockchain-based billing system [73] whose concept can be applied into the health

insurance billing situation. However, in order to achieve the best results, it is necessary to address the proposed solution's limits first. The authors of this study have deployed rest servers and demonstrated that these can actually accelerate software development; nevertheless, they have not considered server security in order to incorporate higher security for identity management.

In addition to security, user privacy is a particularly delicate issue in any framework that must be addressed prior to using this system for health insurance, as users are highly worried about their health information. The healthcare and health insurance sectors heavily rely on data. However, the nature of health data is so sensitive since it contains personally identifiable information that every individual is exceedingly protective of his or her data. If the system for data sharing is insufficiently secure, privacy breaches can occur, causing serious harm to any individual. Therefore, data privacy and system security are key considerations when offering a solution to health insurance problems.

The authors [65] have presented a way for addressing complaints and concerns in the health insurance business. However, they have neglected to account for user privacy, a crucial component of any complaint and grievance management system. On any online forum for complaint and grievance management, if the user's identity is not anonymous, the majority of users do not feel comfortable sharing their issues. Similarly, although the authors of [73] have offered a solid strategy that emphasizes the billing notion of health insurance, they have not disclosed how they protect their customers' privacy.

- **Lack of accuracy and paperwork:** The success rate of a mechanism is proportional to its accuracy. Therefore, any proposed method must strive for the best level of accuracy possible. Based on their findings, the anti-fraud method for healthcare insurance presented in [30] is quite reliable. However, the accuracy falls short of expectations. Similarly, EhtInsurance [31] is a potential method for the handling of health insurance claims that has been shown to be reliable and secure. However, it lacks sufficient documentation.
- **Lack of emphasis on key factors:** Prior to implementing a solution, the first step in attempting to handle a problem in the real world is to identify the major components that must be addressed. Without it, obtaining maximum achievement is somewhat unpredictable and difficult. Consequently, the first stage in fixing a problem is to identify the crucial parts that must be addressed prior to implementing the solution. However, several of the research have not adhered to this fundamental guideline, which is a major flaw of their system. ShareChain [82] presents a data sharing structure that is mainly absent of this characteristic.

Numerous studies have emphasized the need of secure and effective data sharing in the health insurance industry. ShareChain [82] is one of them, and it has introduced an efficient data-sharing solution to address interoperability and reliability issues. Existing works have been compared to the proposed framework to evaluate their success and limits. The primary issue with this strategy is that it lacks a patient-centered approach. Due to the fact that the customer or patient is at the heart of the healthcare or health insurance industry and is the ultimate owner of their data, the framework should concentrate largely on them. If data collection fails, it is useless to adopt any data-sharing system, and the insurance industry suffers huge losses.

4.10.2 Future directions

In the previous sub section, the limitations of existing blockchain-based health insurance management systems have been highlighted. Based on these limitations, potential future directions have been provided in this subsection.

- **Need for effective solutions:** There is a large gap in the use of blockchain technology to address several areas of health insurance. As indicated in the preceding section, monitoring and response management are lacking in the literature and are not being investigated by researchers. The authors of [87] have provided a monitoring framework that future researchers might use to generate ideas for health insurance monitoring and response management. Perhaps this remains an open question that future scholars should explore. However, there has been a few research on blockchain-based complaint and grievance management systems and blockchain-based billing management systems in numerous areas [88,89]. Using this work, researchers can attempt to integrate these concepts into the health insurance system by resolving the underlying dependencies.

  Compared to the preceding aspect, numerous contributions have been made to claim management and fraud detection in health insurance. Mentioned are several potential approaches that could be utilized in the context of health insurance claim management and the detection of fraudulent activity. The authors of [67] have suggested a blockchain-based strategy for developing a tamper-proof claims processing system for health insurance claims administration. Detecting fraud and preventing risks in health insurance is another area where blockchain has been utilized in multiple research. This chapter looks at the most promising studies from the authors' point of view to give readers a full understanding of how fraud happens and how to stop it.
- **Solutions' applicability and scalability:** A common behavior among health insurance researchers is a lack of awareness or motivation to verify the applicability and scalability of their suggested solutions. There are a few excellent solutions for various aspects of health insurance, such as a complaint redressal system from [64], an anti-fraud system from company [75], and a claim management system from company [67], but none of the architectures have underwent applicability and scalability testing. Future research should be conducted to demonstrate the efficacy and scalability of their proposed method so that it can be applied to the detection of fraud in health insurance and other aspects.

  Regarding the possibility of incorporating an existing solution from another sector into the health insurance system, two considerations are outlined in Section 4.1. Therefore, substantial research is necessary to comprehend the applicability and scalability of existing health insurance systems.

- **Increasing trend in simulation and benchmarking:** As described in the preceding section, simulation and benchmarking play a crucial role in assessing any architecture. If it is unknown whether a newly offered solution will outperform current alternatives, then there is no reason to choose the new one over the old one. This is where simulation and benchmarking come into play, something the majority of studies have lacked. Consequently, future researchers must incorporate simulation and benchmarking prior to deeming their solution beneficial. Additionally, case studies should be presented in exceptional circumstances, such as a consumer emergency complaint registry.
- **Increase feasibility:** For the successful integration of a certain solution, it is essential to ensure that it is feasible. Several of the works that we analyzed lacked this quality. For instance, Ismail *et al.* have developed a very successful method for claim management in health insurance, however it fails to meet some practical situations. As stated previously, authors set up rules for their component, which is not feasible at all. Setting up rules for every component, especially in fraudulent activities, is an unstable state. Critical scenarios such as medical treatment requiring rigorous validation need a lot of expertise and knowledge. Therefore, additional study must be conducted on the integration of machine learning algorithms and blockchain smart contracts. This will aid in the discovery of fraudulent instances and the overall success of the claims management process.

Likewise, Ismail *et al.* [74] have manually categorized fraudulent acts into twelve distinct groups. However, the amount of possible fraudulent actions remains uncertain. In this regard, substantial research is required so that the behavior of various fraudulent operations can be incorporated into future frameworks. Future additions to this proposed system could include an interoperability framework for data claims.

- **Security:** Numerous studies have the potentiality to overcome health insurance management. But their mechanisms raise security concerns which need utmost consideration. For instance, mechanisms from Thenmozhi *et al.* [67] have not made use of encrypted databases. By utilizing fine-grained access control, this concern can be easily accomplished. Ahmet *et al.* [73] have utilized restful servers, which have introduced the topic of finding better servers. This topic needs exhaustive research in order to integrate greater security for identity management. Additionally, post-quantum cryptography can be investigated and included into the process as quantum computers become more widespread as well as a symmetric encryption should be investigated as an alternative.
- **Privacy:** User privacy is a vital aspect in any complaint and grievance management system. In their complaint and grievance management system, Rahman *et al.* [65] have not considered the user's privacy. Therefore, the communication must be made anonymous by utilizing the blockchain's anonymity feature, as this makes the communication more trustworthy, faster, and safe. Besides, Ahmet *et al.* [73] have made no mention of how their proposed billing system would protect their customers' privacy. Thus, planning to

integrate a new privacy mechanism is a matter of urgency in these cases because health data is continuous in nature. Privacy techniques such as differential privacy are an excellent strategy for ensuring users' data privacy [83].

- **Increase accuracy and paperwork:** For any proposed method to be effective, a minimum level of precision and essential documentation are required. An anti-fraud system proposed by [30] is a very promising solution to the problem of fraud in the health insurance system, but it lacks a crucial element: accuracy. Subsequently, if additional work is done to improve the system's accuracy, this proposed model could be nearly unbeatable. In addition, written documentation permits us to monitor the development of any project and reveals how to enhance the system. Similarly, EhrInsurance [31] is a promising approach to claim management; however, a major element it lacks is proper amount of paperwork.
- **Emphasize on key factors:** The first and most important stage in proposing a novel solution is to identify the key aspects that must be considered. If this is not ensured, there is a significant chance that the proposed method will become a useless solution. ShareChain [82] is a really promising piece of work that focuses on data sharing in the context of health insurance. However, the most critical issue with this solution is that they have not considered making it patient-centered, which is the most important requirement from a health insurance standpoint. Therefore, it must be taken into account for the system to be functional. The authors have also mentioned that they want to put their system in the real world, which is the ultimate objective of any proposal for a data-sharing platform.

4.11 Conclusions

The benefits of blockchain technology in the health insurance industry are evident. With its core qualities of decentralization, persistence, anonymity, and verifiability, blockchain technology has the potential to influence the existing health insurance sector. This chapter discusses numerous aspects of health insurance and how they might interact using blockchain technology. This emerging technology has the ability to minimize the costs associated with a decentralized environment if employed effectively. Numerous studies have examined insurance claims, fraud detection, and data sharing among insurance participants while preserving their privacy and security. Whereas there is a shortage of literature regarding the proper handling of medical bills for insurance claims and the management of complaints within the insurance cycle. Overall, there are very few studies that have explored the various aspects and challenges of health insurance. In addition, future research ought to concentrate on implementing the blockchain-based service in a real world environment in order to assess the practicality and scalability of such a system. Finally, blockchain technology has the ability to alleviate health insurance concerns through decentralization and the elimination of third parties, thereby accelerating the process. Regarding blockchain, health insurers must be fearless. To utilize the unique characteristics of blockchain technology, insurer have to have the courage to develop a new security policy, strengthen the entire process, and contribute to the company's success.

References

[1] Keisler-Starkey K, Bunch LN. Health Insurance Coverage in the United States: 2020. United States Census Bureau. 2021.

[2] KAGAN J. Health Insurance; Updated March 06, 2022. [Online; accessed April 7, 2022]. https://www.investopedia.com/terms/h/healthinsurance.asp.

[3] What is Health Insurance? [Online; accessed April 7, 2022]. https://www.iciciprulife.com/health-insurance/what-is-health-insurance.html.

[4] Health Insurance – Meaning & Definition; Updated November 04, 2016. [Online; accessed April 7, 2022]. https://www.hdfclife.com/insurance-knowledge-centre/about-life-insurance/health-insurance-meaning-and-types.

[5] Mendoza-Tello JC, Mendoza-Tello T, Villacs-Ramón J. A blockchain-based approach for issuing health insurance contracts and claims. In: *The International Conference on Advances in Emerging Trends and Technologies*. New York, NY: Springer; 2021. p. 250–260.

[6] Levit LA, Balogh E, Nass SJ, *et al. Delivering High-Quality Cancer Care: Charting a New Course for a System in Crisis*. National Academies Press Washington, DC; 2013.

[7] Ellis L, Canchola AJ, Spiegel D, *et al.* Trends in cancer survival by health insurance status in California from 1997 to 2014. *JAMA Oncology*. 2018; 4(3):317–323.

[8] Walker GV, Grant SR, Guadagnolo BA, *et al.* Disparities in stage at diagnosis, treatment, and survival in nonelderly adult patients with cancer according to insurance status. *Journal of Clinical Oncology*. 2014; 32(28):3118.

[9] Ayanian JZ. America's Uninsured Crisis: Consequences for health and health care. Statement before the Committee on Ways and Means, United States House of Representatives Public Hearing on Health Reform in the 21st Century: Expanding Coverage, Improving Quality and Controlling Costs March. 2009. p. 11.

[10] Halpern MT, Ward EM, Pavluck AL, *et al.* Association of insurance status and ethnicity with cancer stage at diagnosis for 12 cancer sites: a retrospective analysis. *The Lancet Oncology*. 2008;9(3):222–231.

[11] Ward E, Halpern M, Schrag N, *et al.* Association of insurance with cancer care utilization and outcomes. *CA: A Cancer Journal for Clinicians*. 2008; 58(1):9–31.

[12] Mandelblatt JS, Yabroff KR, Kerner JF. Equitable access to cancer services: a review of barriers to quality care. *Cancer: Interdisciplinary International Journal of the American Cancer Society*. 1999;86(11):2378–2390.

[13] Zamosky L. Healthcare, Insurance, and You: The Savvy Consumer's Guide. Apress; 2013.

[14] Sommers BD, Gourevitch R, Maylone B, *et al.* Insurance churning rates for low-income adults under health reform: lower than expected but still harmful for many. *Health Affairs.* 2016;35(10):1816–1824.

[15] Health Insurance – Meaning & Definition; Updated November 04, 2016. [Online; accessed April 7, 2022]. https://www.hdfclife.com/insurance-knowledge-centre/about-life-insurance/health-insurance-meaning-and-types.

[16] Chu V. National Health Care Fraud and Opioid Takedown Results; Updated September 30, 2020. [Online; accessed April 7, 2022]. https://www.justice.gov/usao-sdca/pr/national-health-care-fraud-and-opioid-takedown-results-charges-against-345-defendants.

[17] Association NHCAF, *et al. Cost of healthcare frauds.* Accessed: March 2021. p. 11.

[18] Donovan F. Half of US Adults Are Anxious About Healthcare Data Security; July 30, 2018. [Online; accessed April 10, 2022]. https://healthitsecurity.com/news/half-of-us-adults-are-anxious-about-healthcare-data-security.

[19] Kose I, Gokturk M, Kilic K. An interactive machine-learning-based electronic fraud and abuse detection system in healthcare insurance. *Applied Soft Computing.* 2015;36:283–299.

[20] Roy R, George KT. Detecting insurance claims fraud using machine learning techniques. In: *2017 International Conference on Circuit, Power and Computing Technologies (ICCPCT).* New York, NY: IEEE; 2017. p. 1–6.

[21] Branting LK, Reeder F, Gold J, *et al.* Graph analytics for healthcare fraud risk estimation. In: *2016 IEEE/ACM International Conference on Advances in Social Networks Analysis and Mining (ASONAM).* New York, NY: IEEE; 2016. p. 845–851.

[22] Srinivasan U, Arunasalam B. Leveraging big data analytics to reduce healthcare costs. *IT Professional.* 2013;15(6):21–28.

[23] Selvakumar V, Satpathi D, Praveen Kumar P, *et al.* Modeling and prediction of third-party claim using a machine learning approach. *Indian Journal of Science and Technology.* 2020;13(21):2071–2079.

[24] IBM. What is blockchain technology?. [Online; accessed April 7, 2022]. https://www.ibm.com/topics/what-is-blockchain.

[25] Namasudra S, Deka GC, Johri P, *et al.* The revolution of blockchain: state-of-the-art and research challenges. *Archives of Computational Methods in Engineering.* 2021;28(3):1497–1515.

[26] HAYES A. Blockchain Explained; Updated March 05, 2022. [Online; accessed April 7, 2022]. https://www.investopedia.com/terms/b/blockchain.asp.

[27] MORRIS DZ. Leaderless, Blockchain-Based Venture Capital Fund Raises $100 Million, And Counting; May 16, 2016. [Online; accessed April 11, 2022]. https://fortune.com/2016/05/15/leaderless-blockchain-vc-fund/.

[28] Economist T. A Venture Fund With Plenty of Virtual Capital, but No Capitalist; October 31, 2015. [Online; accessed April 11, 2022]. https://innovationtoronto.com/2016/05/venture-fund-plenty-virtual-capital-no-capitalist/.

[29] Chowdhury MJM, Usman M, Ferdous MS, *et al.* A cross-layer trust-based consensus protocol for peer-to-peer energy trading using fuzzy logic. *IEEE Internet of Things Journal.* 2021;99:1.

[30] Liu W, Yu Q, Li Z, *et al.* A blockchain-based system for anti-fraud of healthcare insurance. In: *2019 IEEE 5th International Conference on Computer and Communications (ICCC).* New York, NY: IEEE; 2019. p. 1264–1268.

[31] Sawalka S, Lahiri A, Saveetha D. EthInsurance: a blockchain based alternative approach for Health Insurance Claim. In: *2022 International Conference on Computer Communication and Informatics (ICCCI)*; 2022. p. 1–9.

[32] Contributors E. Overview: Insurance. [Online; accessed April 12, 2022]. https://www.encyclopedia.com/finance/encyclopedias-almanacs-transcripts-and-maps/overview-insurance.

[33] Hamer L. Here's How Health Insurance Companies Ripped Off People For Years; January 15, 2018. [Online; accessed April 12, 2022]. https://www.amazon.com/Bitcoin-Cryptocurrency-Technologies-Comprehensive-Introduction/dp/0691171696.

[34] Cutler DM, Wikler E, Basch P. Reducing administrative costs and improving the health care system. *New England Journal of Medicine*. 2012; 367 (20):1875–1878.

[35] Weinick RM, Burns RM, Mehrotra A. Many emergency department visits could be managed at urgent care centers and retail clinics. *Health Affairs.* 2010;29(9):1630–1636.

[36] Vogeli C, Shields AE, Lee TA, *et al.* Multiple chronic conditions: prevalence, health consequences, and implications for quality, care management, and costs. *Journal of General Internal Medicine.* 2007;22(3):391–395.

[37] Ayanian JZ, Williams RA. Principles for eliminating racial and ethnic disparities in healthcare. In: *Eliminating Healthcare Disparities in America*. New York, NY: Springer; 2007. p. 377–389.

[38] Freeman JD, Kadiyala S, Bell JF, *et al.* The causal effect of health insurance on utilization and outcomes in adults: a systematic review of US studies. *Medical Care.* 2008;p. 1023–1032.

[39] Goodnough A. Census shows a modest rise in US income. *New York Times A*. 2007;1.

[40] Institute of Medicine (US) Committee on the Consequences of Uninsurance I. Care Without Coverage: Too little, Too Late. Washington, DC: National Academy Press; 2002.

[41] Bush G, Carter J, Ford G, *et al. Building a Better Health Care System: Specifications for Reform*. Washington, DC: National Coalition on Health Care; 2004.

[42] Institute of Physicians-American Society of Internal Medicine AC, *et al. No Health Insurance? Its Enough to Make You Sick: Scientific Research Linking the Lack of Health Coverage to Poor Health*. Philadelphia, PA: American College of Physicians-American Society of Internal Medicine. 1999.

[43] Brown ME, Bindman AB, Lurie N. Monitoring the consequences of uninsurance: a review of methodologies. *Medical Care Research and Review*. 1998;55(2):177–210.

[44] Sicker HJ. Sicker and poorer—The consequences of being uninsured: a review of the research on the relationship between health insurance, medical care use, health, work, and income. *Medical Care Research and Review*. 2003;60(2 Suppl):3S–75S.

[45] Hoffman C, Paradise J. Health insurance and access to health care in the United States. *Annals of the New York Academy of Sciences*. 2008;1136 (1):149–160.

[46] Howell EM. The impact of the Medicaid expansions for pregnant women: a synthesis of the evidence. *Medical Care Research and Review*. 2001;58 (1):3–30.

[47] Kilbourne AM. Care without coverage: too little, too late. *Journal of the National Medical Association*. 2005;97(11):1578.

[48] Congress U. Office of Technology Assessment: Does Health Insurance Make a Difference. In: *Background Paper*. Washington, DC: US Congress; 1992.

[49] Weissman JS, Epstein AM. The insurance gap: does it make a difference? *Annual Review of Public Health*. 1993;14:243–270.

[50] MarketsandMarkets. Blockchain In Insurance Market by Provider, Application (GRC Management, Death & Claims Management, Identity Management & Fraud Detection, Payments, and Smart Contracts), Organization Size (Large Enterprises and SMEs), and Region – Global Forecast to 2023; July 2018. [Online; accessed April 13, 2022]. https://www.marketsandmarkets.com/Market-Reports/blockchain-in-insurance-market-9714723.html.

[51] Chirag. How Blockchain Technology is Transforming the Insurance Industry; December 16, 2021. [Online; accessed April 13, 2022]. https://appinventiv.com/blog/blockchain-transforming-the-insurance-industry/.

[52] Review AI. How blockchain could revolutionise the insurance industry; Jun 2018. [Online; accessed April 13, 2022]. https://appinventiv.com/blog/blockchain-transforming-the-insurance-industry/.

[53] Amponsah AA, Weyori BA, Adekoya AF. Blockchain in insurance: exploratory analysis of prospects and threats. *International Journal of Advanced Computer Science and Applications*. 2021;12(1):445–466.

[54] Borah MD, Visconti RM, Deka GC. *Blockchain in Digital Healthcare. London:* CRC Press; 2021. Available from: https://books.google.com.bd/books?id=vQdUEAAAQBAJ.

[55] Kuo TT, Kim HE, Ohno-Machado L. Blockchain distributed ledger technologies for biomedical and health care applications. *Journal of the American Medical Informatics Association*. 2017;24(6):1211–1220.

[56] Cheung CF, Lee WB, Wang WM, *et al.* A multi-perspective knowledge-based system for customer service management. *Expert Systems with Applications*. 2003;24(4):457–470. Available from: https://www.sciencedirect.com/science/article/pii/S0957417402001938.

[57] Vickery SK, Jayaram J, Droge C, *et al.* The effects of an integrative supply chain strategy on customer service and financial performance: an analysis of direct versus indirect relationships. *Journal of Operations Management*. 2003;21(5):523–539. Available from: https://www.sciencedirect.com/science/article/pii/S0272696303000627.

[58] Li Z, Guo H, Wang WM, *et al.* A blockchain and AutoML approach for open and automated customer service. *IEEE Transactions on Industrial Informatics*. 2019;15(6):3642–3651.

[59] Wendel S, de Jong JD, Curfs EC. Consumer evaluation of complaint handling in the Dutch health insurance market. *BMC Health Services Research*. 2011;11(1):1–9.

[60] Bendall-Lyon D, Powers TL. The role of complaint management in the service recovery process. *The Joint Commission Journal on Quality Improvement*. 2001;27(5):278–286.

[61] Alkhateeb YM. Blockchain implications in the management of patient complaints in healthcare. *Journal of Information Security*. 2021;12(3):212–223.

[62] Reitsma-van Rooijen M, Brabers A, de Jong J. Bijna 8% wisselt van zorgverzekeraar. Premie is de belangrijkste reden om te wisselen. NIVEL Utrecht; 2011.

[63] Malhotra S, Patnaik I, Roy S, *et al.* Fair play in Indian health insurance. Available at SSRN 3179354. 2018;.

[64] Jattan S, Kumar V, R A, *et al.* Smart complaint redressal system using ethereum blockchain. In: *2020 IEEE International Conference on Distributed Computing, VLSI, Electrical Circuits and Robotics (DISCOVER)*; 2020. p. 224–229.

[65] Rahman M, Azam MM, Chowdhury FS. An anonymity and interaction supported complaint platform based on blockchain technology for national and social welfare. In: *2021 International Conference on Electronics, Communications and Information Technology (ICECIT)*; 2021. p. 1–8.

[66] Namasudra S, Sharma P, Crespo RG, *et al.* Blockchain-based medical certificate generation and verification for IoT-based healthcare systems. *IEEE Consumer Electronics Magazine*. 2023;12(2):83–93.

[67] Thenmozhi M, Dhanalakshmi R, Geetha S, *et al.* Implementing blockchain technologies for health insurance claim processing in hospitals. In: *Materials Today: Proceedings*. 2021. Available from: https://www.sciencedirect.com/science/article/pii/S2214785321019301.

[68] Chowdhury MJM, Chakraborty NR. Captcha based on human cognitive factor. arXiv preprint arXiv:13127444. 2013.

[69] Alex DelVecchio KL. revenue cycle management (RCM); February 2017. [Online; accessed April 13, 2022]. https://www.techtarget.com/searchhealthit/definition/revenue-cycle-management-RCM.

[70] Wikipedia contributors. Medical billing—Wikipedia, The Free Encyclopedia; 2022. [Online; accessed April 9, 2022]. https://en.wikipedia.org/w/index.php?title=Medical_billing&oldid=1071238649.

[71] He X, Alqahtani S, Gamble R. Toward privacy-assured health insurance claims. In: *2018 IEEE International Conference on Internet of Things (iThings) and IEEE Green Computing and Communications (GreenCom) and IEEE Cyber, Physical and Social Computing (CPSCom) and IEEE Smart Data (SmartData)*. New York, NY: IEEE; 2018. p. 1634–1641.

[72] Rivkin JW, Roberto M, Gulati R. Federal Bureau of Investigation, TN - 2001, 2007, and 2009. Harvard Business School Teaching Note 711–487, 2011.

[73] Gür AÖ, Öksüzer S, Karaarslan E. Blockchain based metering and billing system proposal with privacy protection for the electric network. In: *2019 7th International Istanbul Smart Grids and Cities Congress and Fair (ICSG)*. New York, NY: IEEE; 2019. p. 204–208.

[74] Ismail L, Zeadally S. Healthcare insurance frauds: taxonomy and blockchain-based detection framework (block-HI). *IT Professional*. 2021;23 (4):36–43.

[75] Alhasan B, Qatawneh M, Almobaideen W. Blockchain technology for preventing counterfeit in health insurance. In: *2021 International Conference on Information Technology (ICIT)*. New York, NY: IEEE; 2021. p. 935–941.

[76] Mangan D. Health Insurance Paperwork Wastes $375 Billion; January 13, 2015. [Online; accessed April 6, 2022]. https://www.cnbc.com/2015/01/13/health-insurance-paperwork-wastes-375-billion.html.

[77] Bush J, Sandridge L, Treadway C, *et al.* Medicare fraud, waste and abuse. 2017.

[78] Walsh L, Kealy A, Loane J, *et al.* Inferring health metrics from ambient smart home data. In: *2014 IEEE International Conference on Bioinformatics and Biomedicine (BIBM)*. New York, NY: IEEE; 2014. p. 27–32.

[79] Lokhande S, Mukadam S, Chikane M, *et al.* Enhanced data sharing with blockchain in healthcare. In: *ICCCE 2019*. New York, NY: Springer; 2020. p. 277–283.

[80] Hasan IM, Ghani RF. Blockchain for authorized access of health insurance IoT system. *Iraqi Journal of Computers, Communications, Control and Systems Engineering*. 2021;21(3):76–88.

[81] Liang X, Zhao J, Shetty S, *et al.* Integrating blockchain for data sharing and collaboration in mobile healthcare applications. In: *2017 IEEE 28th Annual International Symposium on Personal, Indoor, and Mobile Radio Communications (PIMRC)*. New York, NY: IEEE; 2017. p. 1–5.

[82] Lee AR, Kim MG, Kim IK. SHAREChain: healthcare data sharing framework using Blockchain-registry and FHIR. In: *2019 IEEE International Conference on Bioinformatics and Biomedicine (BIBM)*. New York, NY: IEEE; 2019. p. 1087–1090.

[83] Saifuzzaman M, Ananna TN, Chowdhury MJM, *et al.* A systematic literature review on wearable health data publishing under differential privacy. *International Journal of Information Security*. 2022;21:1–26.

[84] Kish LJ, Topol EJ. Unpatients—why patients should own their medical data. *Nature Biotechnology*. 2015;33(9):921–924.

[85] Mangaonkar OA, Shah D. Health Insurance management process in hospitals using blockchain secured framework. *International Journal of Research in Engineering, Science and Management*. 2021;4(10):77–79. Available from: http://www.journals.resaim.com/ijresm/article/view/1435.

[86] Goyal A, Elhence A, Chamola V, *et al.* A blockchain and machine learning based framework for efficient health insurance management. In: *Proceedings of the 19th ACM Conference on Embedded Networked Sensor Systems. SenSys '21*. New York, NY: Association for Computing Machinery; 2021. p. 511–515. Available from: https://doi.org/10.1145/3485730.3493685.

[87] Agrawal D, Minocha S, Namasudra S, *et al.* A robust drug recall supply chain management system using hyperledger blockchain ecosystem. *Computers in Biology and Medicine*. 2022;140:105100.

[88] Hingorani I, Khara R, Pomendkar D, *et al.* Police complaint management system using blockchain technology. In: *2020 3rd International Conference on Intelligent Sustainable Systems (ICISS)*. New York, NY: IEEE; 2020. p. 1214–1219.

[89] Yaqub R, Ahmad S, Ali H, *et al.* AI and blockchain integrated billing architecture for charging the roaming electric vehicles. *IoT*. 2020;1(2):382–397.

Chapter 5

e-Healthcare data security using blockchain technology

Amiya Karmakar[1], Pritam Ghosh[2], Partha Sarathi Banerjee[3] and Debashis De[1]

Abstract

The current medical services suffer from critical difficulties like detailed information, detectability, unchanging nature, review, information provenance, universal access, trust, protection, and security. Additionally, an enormous piece of the existing medical services framework utilized for overseeing data is brought together, which might suffer from a single point of failure in cataclysmic events. This limitation must be overcome using advanced decentralized, secure technologies like blockchain. Blockchain is a decentralized technology that can alter, reshape, and change how information is being taken care of in the medical sector. This chapter talks about utilizing blockchain to share medical data securely. This chapter presents the critical blockchain elements and attributes from a security perspective. It also examines the benefits of blockchain innovation alongside potential open doors for the medical sector to preserve privacy and security. At last, this chapter presents ongoing tasks and contextual investigations to show the reasonableness of blockchain innovation for different healthcare applications and areas. Furthermore, this proposed work distinguishes and talks about significant open examination challenges upsetting the fruitful reception of blockchain in the healthcare area.

Keywords: Blockchain; IoT; e-Healthcare; Quantum identity; Security; NFT

5.1 Introduction

Healthcare is one of the areas where blockchain is seen to have enormous promise. Industry can use blockchain innovation to help drug solutions and inventory networks. Supplier accreditations, clinical charging, contracts, clinical record trading,

[1]Centre of Mobile Cloud Computing, Department of Computer Science and Engineering, Maulana Abul Kalam Azad University of Technology, West Bengal, India
[2]Department of Computer Science and Technology, Iswar Chandra Vidyasagar Polytechnic, West Bengal, India
[3]Department of Information Technology, Kalyani Government Engineering College, West Bengal, India

clinical preliminaries, and other areas of healthcare can benefit from blockchain innovation. Healthcare is evolving to support a patient-centered approach [1]. Patients would have control over their medical data. Therefore blockchain-based medical services frameworks might improve security and consistency. Those frameworks could likewise assist with uniting patient information, empowering the trading of clinical records across various healthcare foundations [2].

The transparency and communication between patients and healthcare professionals have improved because of the introduction of blockchain technology in the sector. Due to redundancies, the use of various names and identities, and their availability across many networks, the amount and complexity of healthcare records are expanding but have not yet been optimized [3]. Additionally, it is now crucial to maintain data security and stop the illicit activity. Patient data can be utilized or sold if unauthorized people are permitted access, and everyone with access will be able to see the patients' personal information. To effectively administer healthcare, patient data privacy is essential [4,5]. Utilizing blockchain technology enabled by Industry 4.0 can address these challenges and concerns by ensuring data integrity and preventing manipulation and failure at any one point.

In the current system, businesses' main worries about information sharing between various institutions are security and trust. There are trust concerns as a result of information entering at any point along the communication chain, particularly in the healthcare sector. Concerns exist when the same patient record is held by multiple vendors in different versions that have not been validated, leading to various mistakes, inconsistencies, and incompleteness. It is hardly a surprise that healthcare professionals are worried when tales of security breaches, the tampering of personal data, and the constant threat of hacking are added.

Blockchain technology may provide a solution to the majority of these issues because it is cryptographically secure and allows for the authentication of data using digital signatures that are personal to each user.

While society tends toward getting into a data-driven society's contemplations, data protection issues are also increasing. Centralized systems hoard much individual information and sensitive data [6,7]. When generating enormous amounts of data, people have no control over their data as supersensitive information like patient wellbeing records should not be shared with third parties. As a result, they are inclined to misuse the data [3].

This chapter provides a thorough audit and analysis of blockchain research on the subject of Health Care 4.0. This chapter aims to demonstrate the projected use of security in Blockchain for healthcare and comprehend blockchain-based research's challenges and potential future directions. This proposed work incorporates research that exhibits algorithms, techniques, procedures, or designs for the field of healthcare. In addition, this chapter addresses the accompanying research questions:

- Concentrate on the security of blockchain innovation and its critical necessities in medical services.
- Distinguish capacities of Blockchain-based Healthcare Data Storage and Security.
- To examine the Security of e-Healthcare Data using Blockchain Technology.
- To distinguish and analyze the critical utilization of blockchain for medical services.

5.1.1 Motivation

Important policy choices, patient care, and medical diagnostics, to mention a few, all depend on healthcare data. The significance and commercial demand of healthcare data make it a target for cyberattacks. The centralized record-keeping methods make one node available for assault by attackers. The patient is kept at the center of a decentralized system, which is computationally expensive but potentially revolutionary since it offers security, transparency, privacy, and interoperability of electronic healthcare data. A blockchain implements such a distributed and decentralized system employing trustworthy cryptographic techniques. In-depth audits and analyses of blockchain research on Health Care 4.0 are provided in this publication. Blockchain technology is a byproduct of contemporary society's efforts to meet needs in several healthcare-related applications. Blockchain technology may significantly improve the patient experience while maintaining system security goals. In order to facilitate future debate, the many drivers and benefits of implementing blockchain technology in healthcare are investigated, recognized, and organized into categories based on a study of the research.

This chapter seeks to explain the due application of security in blockchain technology for healthcare and to understand the difficulties and potential future directions in blockchain-based research. Research that demonstrates algorithms, techniques, processes, or designs for the healthcare industry is incorporated into the proposed work.

5.1.2 Contributions

The significant contributions of this chapter are as follows:

1. A brief overview of the security of blockchain-based application in the healthcare industry is primarily described.
2. Benefits and challenges of blockchain are elaborated concisely for the healthcare industry.
3. A brief idea about blockchain and its application in healthcare 4.0 has been provided.
4. In this work, various blockchain applications and use cases and applications are established.

5.1.3 Organization of the chapter

The rest of the chapter is organized as follows. Section 5.2 examines Healthcare Data Storage and Security: A Conceptual Scenario. Section 5.3 presents what Blockchain is. Section 5.4 describes the Smart Contract modeling, and Section 5.5 discusses Improving the Security of e-Healthcare Data using Blockchain Technology. Section 5.6 expressed Recording Patient Consent using NFT. Section 5.7 presented Security Breaches to using Blockchain Technology in Healthcare Industry. In Section 5.8, future scope in the healthcare industry is explained. Finally, in Section 5.9, concluding remarks of this chapter is explained.

5.2 Healthcare data storage and security: a conceptual scenario

This section analyzes the enhancement of technologies in healthcare and the vital security prerequisites for execution in the healthcare industry. An outline of blockchain innovation is presented, including its advantages to medical services frameworks [8].

5.2.1 Evolution of the healthcare industry

During the 1970s, the rise of pre-automated IT frameworks was noticed, called healthcare 1.0. In this period, medical services frameworks were restricted and not facilitated with advanced frameworks because of the absence of new technologies. Also, bio-clinical types of machinery were not yet grown and did not incorporate into arranged electronic gadgets. In this period, only paper-based reports and solutions were generally utilized in healthcare associations, increasing the costs and time associated with healthcare [9–11]. Healthcare 2.0 is based on the opportunities for changing medical services, which began with the presentation of eHealth in 1990. Early instances of Health 2.0 were the utilization of a particular arrangement of Web devices by entities in healthcare, including specialists, patients, and researchers, utilizing standards of open source and client-produced content. The force of organizations and informal communities to customize medical services [12,13]. The coming of healthcare 3.0 matched the idea of Web 3.0, empowering client integration of how easily healthcare data were conveyed. UIs became more straightforward and customizable, taking advanced and customized encounters into account. Electronic healthcare records (EHRs) were additionally presented, alongside wearable and implantable frameworks, empowering constant, pervasive following of patients' medical services. Likewise, EHR frameworks started to rise, incorporating independent non-arranged frameworks, including web-based media. It helps store consumer records; this enables the distribution of healthcare information over-organized secure channels, with online media, or between clinicians, utilizing less complex EHR frameworks; cooperation and correspondence between healthcare suppliers were also improved [14–16]. Healthcare 4.0 (H4.0) adjusts standards and applications from the industry 4.0 development to medical services, empowering endless customization of care to patients and experts. H4.0 might conceivably uphold strong execution in healthcare frameworks, which alludes to their versatile ability to adapt to intricacy. This chapter investigates the effect of ten H4.0 advanced innovations on four capacities of universal frameworks (screen, expect, answer, and learn) concerning emergency clinics. From 2016 to the current day, organizations have encountered healthcare 4.0. Industry 4.0 discusses the ideas where Hi-contact and Hi-tech frameworks are being presented, utilizing distributed computing, haze, and edge computing to fabricate blockchains for helping ongoing admittance to patients' clinical data. The primary point of this period is to improve virtualization, empowering customized healthcare in real time. The emphasis is on cooperation, rationality, and combination, making healthcare prescient and customized [17–19]. With the coming of enormous data, the size and the intricacy of healthcare data are expanding. Documents are frequently copied, crisscrossed applying several naming methods, and are made available on multiple organizations and registries of medical services frameworks. Security is turning out to be increasingly significant for protecting data from security breaches and crimes in healthcare. Assuming unapproved clients can get contact with patient information. The third-party may sell or distribute in the open marketplace, with patients' data being uncovered to anybody with access to the Internet. This data might incorporate addresses, phone numbers, complete names, etc. Thus, the security of patients' information is fundamental in healthcare 4.0. Different nations have proposed or made controlled principles for healthcare frameworks to forestall digital dangers, which further develops a classification of patient data and trust in the supplier–patient relationship. As of now, most healthcare frameworks utilize incorporated client–server-based designs; a focal authority has full admittance to the framework. In this situation, a lack of protection or security defects might prompt breaches in the framework [20–22]. The evolution of the healthcare industry is presented in Figure 5.1.

5.2.2 Common threats in the healthcare industry

Medical data has progressed from paper-based management to computerized information over the last four decades. However, many forms of medical data have yet to be digitized or linked to Health Data Management systems. When considering data vulnerability in healthcare, people can ask, "How sensitive are personal medical records?" Moreover, the solution is simple—supersensitive. There have been numerous security incidents in healthcare systems. Cyber-attacks are a significant source of concern in the healthcare sector since they can quickly jeopardize the cybersecurity of systems and the health and safety of people.

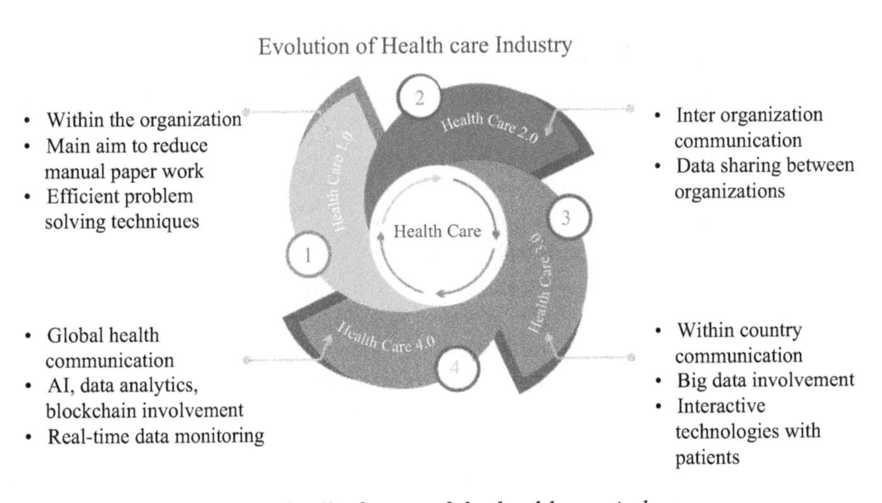

Figure 5.1 Evolution of the healthcare industry

5.2.3 Threat model

The healthcare sector is a top target for cybercriminals because of lax cybersecurity procedures, sensitive data storage, and a desperate need to maintain business continuity. The four major cybersecurity concerns facing the healthcare sector are outlined below to demonstrate the significance of healthcare cybersecurity programs in the context of recent cyberattacks. The security of healthcare data and patient information is most dangerous from these cyberthreats. The most common cybersecurity threat in healthcare is phishing. Phishing is the act of inserting harmful links into a seemingly innocent e-mail. E-mail phishing is the most typical kind of phishing. Phishing e-mails sometimes make reference to a well-known medical condition to encourage link clicking. As a result, they can appear very convincing. Some advanced threat actors create phishing e-mails as replies in an ongoing email conversation to increase authenticity and reduce suspicion. When a person clicks a link in an email scam, they are frequently taken to a fake web page that looks like the login page for well-known corporate software. When these credentials are submitted, fraudsters immediately utilize them to access medical systems. Healthcare information security initiatives should make phishing defenses a point of reference because most cybercrime starts with a phishing attack and has one of the greatest financial effects on an organization. Malware is introduced into a network during a ransomware assault to infect and encrypt sensitive data until a ransom is paid. A phishing assault is typically used to introduce this harmful software into a system. An investigation conducted last year found that ransomware attacks represent an increasing threat to healthcare providers. In 2020, more than one in three healthcare businesses on a global scale were attacked with ransomware. The reason for its prevalence is that hackers know how important it is for the healthcare industry to keep operational disruptions to a minimum. Healthcare victims of ransomware attacks become terrified of the legal repercussions that may follow the theft of patient data. Introducing new technology to automate attacks is another factor contributing to the prevalence of ransomware attacks. Hackers have developed a variant of the Business as a Service (BaaS) model known as Ransomware-as-a-Service, which was inspired by the concept's implementation efficiency (Raas). Traditional technical proficiency is no longer required to launch a ransomware assault under the RaaS paradigm. Similar to how BaaS users do not need to be professionals to become adept in a subject area covered by a BaaS solution, any wannabe cybercriminal may sign up and conduct an attack with the program with minimal cyberattack knowledge. Healthcare IT weaknesses frequently exploited during ransomware attacks must be addressed with the proper security measures to counter the serious security threat posed by ransomware. Comparatively speaking, the healthcare sector has a disproportionately high number of data breaches. HIPAA lays forth stringent guidelines for preventing unauthorized access to sensitive information like health data, but many healthcare organizations have trouble putting its security rules into practice. Despite efforts to limit these occurrences through frameworks like HIPAA, such cybersecurity gaps give cyber attackers access points via which they can continue to compromise the security of medical care data. Such ignored vulnerabilities can be found with an attack surface monitoring solution, extending the work already covered by security budgets without needing a comprehensive cybersecurity makeover. Furthermore, since this threat scenario may make it easier for third parties to gain indirect access to sensitive information like social security numbers, credit card numbers, or even the intellectual property of medical devices, a perfect solution would be able to address third party vendor networks as well. A distributed denial-of-service attack involves flooding a targeted server with bogus connection requests in an effort to take it offline. This coordinated attack uses a large number of endpoints and IoT devices that have been forcibly recruited into a botnet through malware infection. DDoS assaults have the same operational disruption impacts as ransomware attacks but do not carry the same data exfiltration threats. DDoS assaults have the critical advantage that they can cause the same disruption without compromising a network, making them simpler to implement on a much larger scale. Additionally, they have adopted the ransom model due to the speed and destruction that these attacks are capable of. Now, DDoS attackers might take a medical facility offline and only end their attack after receiving a predetermined ransom. Fortunately, a network of Reverse Proxy servers might be used to lessen the effects of these attacks.

5.2.4 How to keep patient data secure

Due to the sensitivity of medical data handled by healthcare providers, they are a possible target for cybersecurity assaults. As a result, system must ensure that data is protected internally and externally. The healthcare business is one of the most tightly managed and vulnerable sectors. As a result, many rules are being implemented to limit personal data access and prevent infractions or breaches. However, the process should begin at its nexus, which is healthcare institutions. Everything from staff communication to a standard for transmitting information among medical apps should be included. In addition, implementing data security methods and safety measures such as data encryption and patient record access restriction to guarantee that nothing is leaked. These include routinely changing passwords, utilizing a virtual private network (VPN), and not keeping data on users' devices. Healthcare data security may be greatly impacted by blockchain as well. All health systems and organizations place a high focus on protecting patient data, but for clinicians, the growing amount of data and management-related concerns provide substantial challenges. Blockchain technology, which employs immutable ledgers that are continuously updated concurrently on all participating network nodes, could be useful in this situation. This indicates that, unlike in a central repository, there is not a single point from which data may be altered. Although offering numerous insecure gateways might also be a concern, blockchain is made to reduce this risk. Blockchains use distinctive signatures or "chains" to link all the data "blocks" that come before and after. Instead of changing the old block when data within a block has to be modified, a new block is introduced to indicate the update. This records every piece of data that is added or changed and includes timestamps.

Blockchains also function via decentralized consensus, which calls for unanimous agreement among all participants in the consortium using the blockchain on the procedures for validating and recording data. For a malicious actor to try to take advantage of this and manipulate the data, they would need to take over most of the network's nodes at once and change the entire blockchain about the data they are aiming for. Although not impossible, this is very challenging because of the extensive network of nodes in the healthcare industry.

5.3 Blockchain architecture

The creation of bitcoin in 2008 popularized blockchain, thanks to a whitepaper written by a mysterious person or group of authors known only as Satoshi Nakamoto. Blockchain is a chain of immutable blocks linked to the preceding root genesis block created by a hierarchy of Distributed Ledger technologies. It keeps track of all transactions that have occurred up to this point. The duplicates of those transactions are updated on all nodes in the decentralized network chain [23]. Before a block is uploaded to the network, each transaction in the block is validated using consensus procedures dispersed across all nodes. Each block is identified with a cryptographic hash, and it mitigates fraud by making the whole linked chain transparent [24]. Private keys are used to sign the transaction, and then it is addressed by the other nodes using the respective public keys, as presented in Figure 5.2.

The core of the blockchain network is balanced by the principles of authentication, non-repudiation, and integrity, which are the main features of asymmetric cryptography. The significance of the number is the success rate of adopting the blockchain by the enthusiasts and the robust security it provides with cryptographic encryptions through digital signatures [25]. Furthermore, another public blockchain called Ethereum can be used for testing transactions, smart contracts, etc. Moreover, blockchains are considered trustless, which means a participant will not have to rely on other users or entities to breach their privacy or not to disrupt the protocols [26]. Types of blockchain and its features are presented in Figure 5.3.

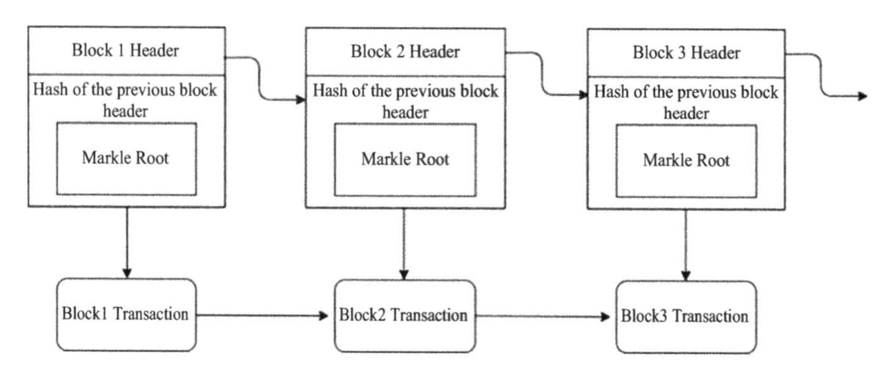

Figure 5.2 Block structure of blockchain

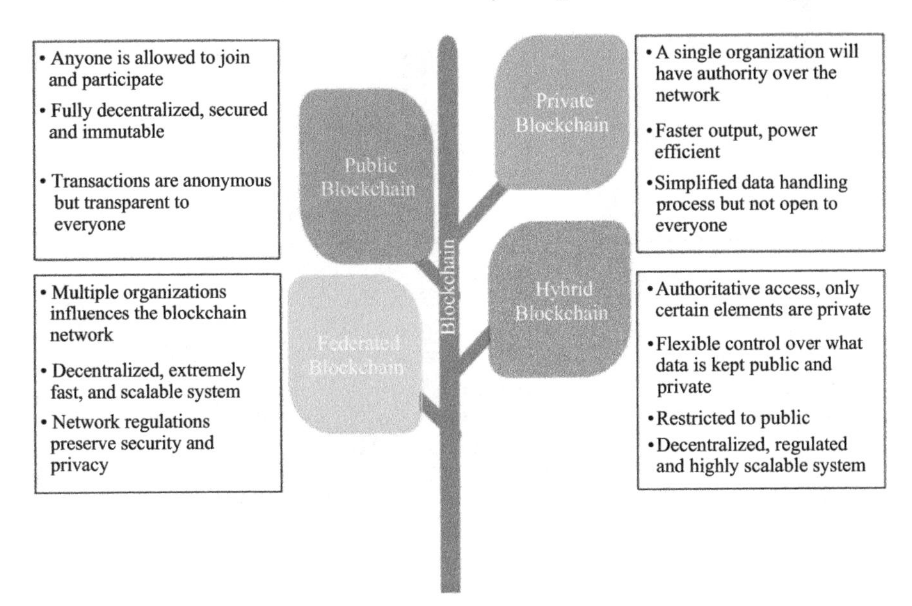

Figure 5.3 Types of blockchain

5.4 Smart contract modeling

A cryptographer named Nick Szabo 1994 proposed the idea of smart contracts [25]. Computers will automatically check this concept of electronic contracts. This revolutionary idea removes centralized entities like banks from the picture. Instead, transactions are self-executing on a distributed trusted network controlled by nodes or computers. A smart contract is a computer code-based agreement between two persons. A smart contract is an agreement between two parties written in computer code. They are stored in a public database and cannot be modified since they run on the blockchain—a distributed ledger. Figure 5.4 shows the components of a smart contract.

Miners must solve a complicated computational arithmetic problem, often known as a "proof-of-work," to gain rewards from verifying transactions using the proof-of-work concept. In plain English, this implies creating a 64-bit hexadecimal integer called a "hash" that is less than or equal to the target hash. It is a gamble, and each step is fraught with chance. To maintain the equilibrium, the more miners competing for a solution, the more complex the challenge becomes, and vice versa. The difficulty of preserving equilibrium decreases as computational capacity is reduced.

5.4.1 Requirement of innovative contract-based blockchain in healthcare 4.0

In this post-COVID-19 world, where medical tests have become a daily norm—a collection of a humongous amount of data is being performed daily. Some of the data may have genetic information, disease history, biological information, food

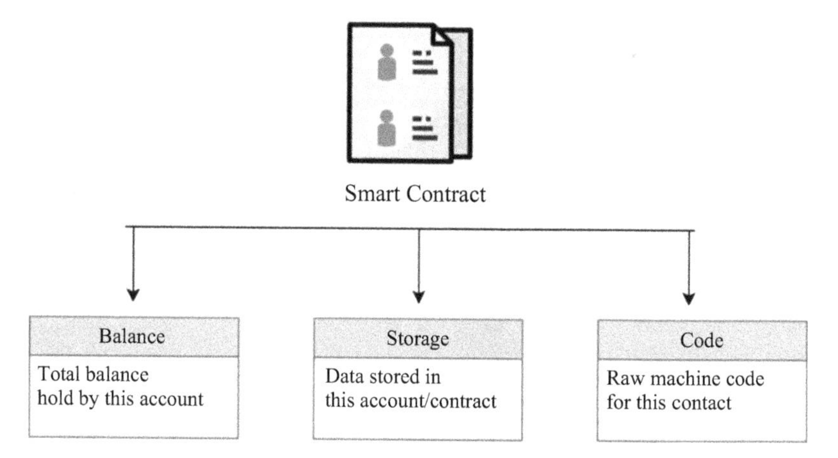

Figure 5.4 Components of smart contract

habits, behavioral habits, and medical reports like MRI, radiology, etc. These are sensitive medical data. To secure the data and prevent leaks, healthcare needs to incorporate next-generation technology like blockchain and artificial intelligence (AI) to minimize the leaks and data inconsistency. It is a known fact by now that IoT is everywhere. The number has reached 50 billion in 2020, and the growth is exponential. Industry can see medical data collection use case specifically in nursing homes or remote medical camps. Much data is collected ad hoc in a monitoring system that is not secure with cutting-edge privacy measures and is highly centralized in moving the data from the remote location to the target data center. If the data is centralized, it may pose a variety of problems. When incorporated with blockchain, a large-scale IoT infrastructure legitimizes the peer-to-peer distributed trustless architecture. Therefore, blockchain is the probable solution to the mentioned issues. The evolution of blockchain is presented in Figure 5.5.

While smart contracts have the potential to help any industry, the healthcare sector can especially benefit from them [27–29]. For example, in a healthcare blockchain transaction, the following happens:

- When a patient interacts with a healthcare organization, a new record of his data, including his age, gender, and address, is created. The patient would be given a private key, and the data would be kept on a blockchain.
- A patient would receive services from the medical facility. The patient's public ID would be used to record information on the procedure, the doctor's note, the medication the pharmacy dispensed, etc.
- The patient and healthcare organization have finished their transaction, which has been uniquely identifiable.
- The patient's public (non-identifiable) ID, which is part of the transaction's identification, is encrypted and given to each transaction, which is then recorded on the blockchain.

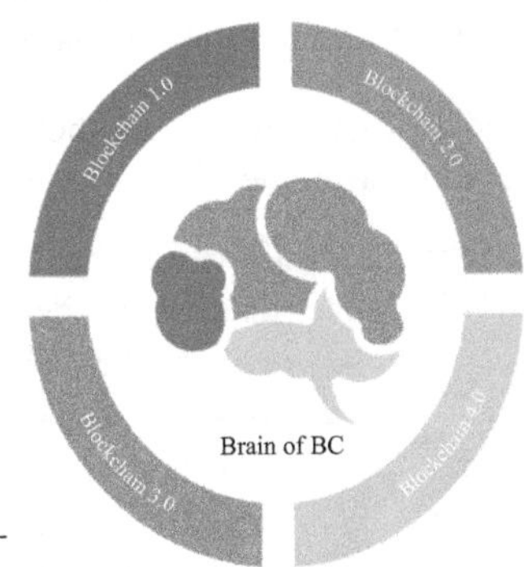

Figure 5.5 Blockchain evolution from 1.0 to 4.0

- In order to get the encrypted data that can be used for medical research, health organizations and universities can submit requests for the data via application programming interfaces (APIs) and use the patient's public (non-identifiable) ID on the blockchain.
- The patient can provide new healthcare organizations his private key so they can use it to decrypt his data. The blockchain data cannot be used to identify a patient without the private key because it links their identity to the data.

5.5 Improving security of e-healthcare data using blockchain technology

According to the specifications for a new version of secure EHR systems [30,31], the main objectives for implementing secure blockchain-based EHR systems are listed below:

- Individual data will only be used secretly, and only those with permission may access the required information.
- Data access is restricted to authorized users only.
- Data must be accurate while in transit and cannot be changed by an unauthorized party (ies).
- Access to information and resources is not unfairly withheld from lawful users.
- Auditability is a crucial element of security. For instance, audit logs primarily record who has accessed which EHR (or specific PHR), why, and the time-stamped details of every action throughout the full life cycle.
- Accountability: a person or a company will be investigated for wrongdoing and held accountably.
- The capacity to verify requestors' identities before granting access to sensitive information.

- For privacy, entities lack any external identifiers. However, it is difficult to be completely anonymous, thus pseudo-anonymity is more widespread (i.e., users are identified by something other than their actual identities).

Current blockchain-based research in the healthcare industry focuses on the following key areas to achieve the aforementioned objectives:

- Data archiving. A variety of confidential healthcare data is kept in a blockchain-based trustworthy ledger database. When secure storage is attained, data privacy should be assured. However, the volume of healthcare data is frequently substantial and intricate. Consequently, a corresponding difficulty is how to handle massive data storage without negatively affecting the functionality of the blockchain network [30,32].
- Data exchange. The primary stewardship of data is often maintained by service providers in the majority of current healthcare systems. However, with the idea of self-sovereignty, there is a movement to give users gain control over their personal data, allowing them to share or not share it as they see fit. Secure data sharing between many companies and domains is also essential [33,34].
- Data analysis. Audit logs can be used when conflicts emerge as evidence to hold requestors responsible for their interactions with EHRs. Some systems use blockchain technology and smart contracts to maintain a trail for auditability. The blockchain ledger will save all transactions and requests, and they can all be retrieved at any time.
- Identification manager. The system must ensure that each user's identity is legitimate. In order to maintain system security and prevent malicious attacks, only authorized users are permitted to submit the necessary requests [35].

5.5.1 Maintaining secured and trusted records of healthcare using blockchain-based federated learning

With the fast advancement of computer science, more and more medical information is generated from clinical establishments, patients, insurance agencies, and drug ventures. This access opens the door to data science advances to determine information-driven insights and improve the model. However, medical care information is fragmented, and it is tough to produce accurate results across health networks. For instance, various emergency clinics own the EHR of various patient populations, and these data are challenging to share across emergency clinics because they are susceptible. This makes a significant boundary for creating an efficient AI model. Federated learning (FL) is a technique of preparing a standard worldwide model with a focal server while storing all the delicate information in local organizations where the data is produced. FL gives an extraordinary guarantee to interface the divided medical services information sources securely. This study aims to provide an insight into privacy-preserving FL, especially inside the biomedical space.

5.5.2 Identity preservation using the blockchain-based application

These days, many clinical gadgets make large sensor networks, specifically, the Internet of Medical Things (IoMT). Therefore, accumulating the information created in various IoMT devices will contribute to an important AI (ML) model, which can be valuable in numerous situations, for example, wellbeing observing, helper determinations, and cataphoretic forecast [36].

However, the well-being-related information in an IoMT usually is firmly connected with individuals' security, which cannot be shared or accumulated nonchalantly. FL is used to address it, which empowers the on-device training without moving the information outside of the device. However, ordinary FL structure depends on a focal server which is helpless against the failure of the focal server. Luckily, the arising blockchain innovation makes a correlation to it. With a blockchain, the accumulation of model informs, and the organization of the training task can be directed securely [37].

The current examinations to join the blockchain and FL predominantly discuss the framework plan and computation advancement, which overlook a fundamental issue of information sparsity in a BFL. For instance, at the point when IoMT comes into administration for the initial time, it can grip meagre information tests in its beginning phase. However, before the information gathering to a considerable sum, the gadgets in the emergency clinic are unequipped for securing recent models or profiting from ML. In addition, if the quantity of IoMT gadgets in the medical clinic is small, it is excessively stretched for the BFL bunch to collect adequate information [38].

An immediate answer for advancing the information tests is to broaden the size of a BFL bunch, which covers as many gadgets as expected under the circumstances. For example, these gadgets might situate in various emergency clinics far away from one another. In this way, network dormancy between devices might be exceptionally high. Then again, the blockchain framework conveyed in the bunch requires incessant organization correspondences to agree. Taken together, came about from the regular and high-dormancy posts, the agreement effectiveness and the comparing framework proficiency can be genuinely low. Additionally, BFL needs the model updates to be dispersed across the FL group.

Therefore, the bigger the group scope, the bigger the likelihood of information protection spillage, mainly when the group is laid out across various clinics. To manage the information sparsity and protection spillage issues while giving high framework productivity, which talks about a cross-group FL structure through the cross-chain method (CFL) in this chapter. Numerous little bunches rather than an enormous group are worked for topographically far-off regions (e.g., medical clinics). In each group, BFL is led with the model updates being accumulated. The accumulated updates are then traded across groups, which improves the updates for each bunch. FL requires continuous correspondence between hubs during the learning system. Consequently, it requires sufficient local power and memory; nonetheless, the innovation likewise maintains a strategic distance from information correspondence, which can require

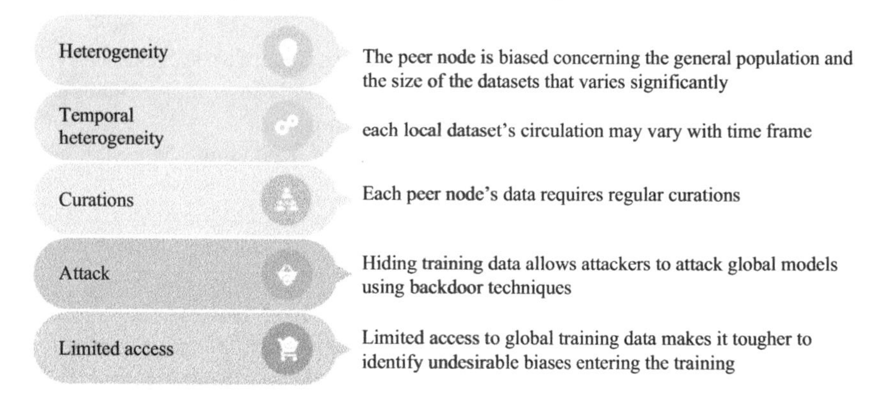

Figure 5.6 Statistical challenges in FL

critical assets before beginning incorporated AI. The devices are integrated into FL, generally communication-constrained, such as IoT devices usually connected to Wi-Fi systems. As a result, even though models are typically less expensive to spread than raw data, FL mechanisms may not be suitable in their general form. FL raises several statistical challenges, as presented in Figure 5.6.

5.6 Recording patient consent using non-fungible tokens

NFT transactions are validated on multiple nodes before being permanently added to the blockchain; NFT ledgers can store an individual's medical records without jeopardizing confidentiality or risking tampering from outside sources. This ensures that every record is accurate and secure from malicious attempts at manipulation. NFT apps have been created specifically to assist healthcare workers, such as NFT Birth Certificates, which healthcare practitioners may issue to infants. Issuing NFTs to each kid may be a fast and easy method to establish a lifetime identity on the blockchain connected to their birth certificate and then validated using NFT verification applications. NFT ledgers also provide a safer way to store sensitive medical information while enabling authorized healthcare practitioners access when needed. Several narrowly defined NFT use cases have emerged in recent years. Hospitals, health insurance companies, and other organizations are exploring how blockchains could help improve hospital operations by verifying patient identities and recording medical procedures without jeopardizing patient confidentiality.

5.7 Security breaches of blockchain in the healthcare industry

Interoperability, security, privacy, scalability, speed, and patient engagement are some of the issues that have been mentioned as obstacles to the development of blockchain-based applications. The lack of a standard for designing blockchain-based healthcare applications poses an interoperability difficulty; as a result, applications produced by different vendors or on different platforms may not be able to communicate with one another. In addition, it would be difficult to transfer data from one platform to the other. Despite the encryption measures used, there is a chance that it may still be feasible to reveal a patient's identity in a public blockchain by putting together sufficient data associated with that patient [39]. Furthermore, there is the possibility of security breaches resulting from harmful attacks on the healthcare blockchain by criminal organizations or government agencies, which might jeopardize patient privacy. For example, several assaults on the blockchain networks that fuel various cryptocurrencies have been reported. In addition, the private keys used in blockchain for data encryption and decryption are similarly vulnerable to compromise, resulting in unauthorized access to recorded health data. The scalability of blockchain-based healthcare systems is a significant hurdle, particularly given the amount of data involved. Storing high-volume biomedical data on the blockchain is not desirable or even feasible in some situations because it will always result in significant performance reduction. There is also the issue of speed, as blockchain-based processing might cause a considerable lag. For example, the Ethereum blockchain platform's current validation mechanism requires all nodes in a network to participate in the validation process. This causes severe processing delays, especially if the data load is large.

5.8 Future scope in the healthcare industry

This section provides insights about upcoming advanced secure technologies and their integration into the healthcare industry. Some of the open research problems of healthcare for future healthcare are as follows.

1. **"Dew-chain" a latency reduction and lightweight blockchain technology in healthcare:** As a Dew-chain integrated healthcare model [40] exploits a significantly decreased latency in healthcare, latency reduction and lightweight node integration is anticipated. It also makes the model Internet-independent and suitable for tiny devices with limited processing power. Furthermore, the dew chain enables the blockchain system to interact with dew devices that can provide an easy user interface for users.
2. **Explainable AI in healthcare:** Explainable AI (XAI) or Interpretable AI is AI that allows people to understand the results of a solution [41]. It defies the notion of a "black box" in machine learning, in which even its creators cannot explain why AI arrived at a particular result. Explainable AI integration in healthcare addresses the current system's decision-making shortcomings. It improves the framework for making intelligent choices based on many environmental variables. This technology helps the user to interact with the system without knowing it's working principle, which removes the complexity of the system.
3. **Quantum identity to reduce unethical digital ethics violations in healthcare:** In the healthcare industry, quantum identity minimizes the difficulties

of digital ethics violations. In healthcare, the Quantum Identification Verification Protocol is a possible way to confirm digital identity through counterintuitive information transfer [42].

4. **Blockchain-based MCDM involvement in healthcare to achieve human perception:** Incorporation of blockchain based multi-criteria decision-making (BMCDM) techniques makes the healthcare model smart enough to make secure decisions based on different aspects of the environment.
5. **Digital twin authentication in healthcare:** Providers can employ digital twins to visually test and assess novel treatment choices on an individual basis. Macro models of hospitals and businesses can direct processes and resource allocation at the organizational level to produce the best, most affordable results.

5.9 Conclusion

Healthcare has a few impediments concerning secrecy, protection, and information honesty. Blockchain and FL are innovations that give inherent security and decentralized means of incorporating AI in healthcare. This work scrutinizes the blockchain to provide healthcare security. Comprehensive study of secure blockchain-based applications and their implementation in the healthcare industry are addressed in this chapter. First, this proposed work presents the advancements in the health industry, blockchain, FL, and NFT innovation. Then, it expands on a few strategies and procedures which empower the blockchain-based healthcare industry in the IoMT space. Finally, further, illustrate the challenges and limitations of blockchain-based technologies are elaborated. In summary, this study demonstrates how blockchain has enormous potential to change the traditional healthcare sector. However, there are still a lot of operational and research difficulties when trying to integrate blockchain technology with current EHR systems completely. This chapter looked over and talked about a few of these difficulties. Next, it identified several possible study areas related to edge computing, big data, machine learning, and IoT. This analysis will help shed more light on the creation and application of the future generation EHR systems, which will be advantageous to our legacy society.

References

[1] Miraz MH. Blockchain of things (BCoT): the fusion of blockchain and IoT technologies. In: *Advanced Applications of Blockchain Technology*. New York, NY: Springer; 2020. p. 141–159.

[2] Hati S and De D. Obsc: osmotic blockchain based framework for smart city environment. In: *2020 Fifth International Conference on Research in Computational Intelligence and Communication Networks (ICRCICN)*. New York, NY: IEEE; 2020. p. 143–148.

[3] He J, Wei J, Chen K, *et al.* Multitier fog computing with large-scale IoT data analytics for smart cities. *IEEE Internet of Things Journal*. 2017;5(2):677–686.

[4] Guha Roy D, Mahato B, Ghosh A, *et al.* Service aware resource management into cloudlets for data offloading towards IoT. *Microsystem Technologies.* 2019;(28):1–15.

[5] Karmakar A, Ganguly K, Ghosh P, *et al.* FemmeBand: a novel IoT application of smart security band implemented using electromyographic sensors based on wireless body area networks. *Innovations in Systems and Software Engineering.* 2022;1–19, DOI: https://doi.org/10.1007/s11334-022-00490-2.

[6] Lu X, Niyato D, Jiang H, *et al.* Ambient backscatter assisted wireless powered communications. *IEEE Wireless Communications.* 2018;25(2):170–177.

[7] Karmakar A, Banerjee PS, De D, *et al.* MedGini: Gini index based sustainable health monitoring system using dew computing. *Medicine in Novel Technology and Devices*. 2022;16:100145.

[8] Dinh TTA, Wang J, Chen G, *et al.* Blockbench: a framework for analyzing private blockchains. In: *Proceedings of the 2017 ACM International Conference on Management of Data*; 2017. p. 1085–1100.

[9] Zyskind G and Nathan O. Decentralizing privacy: using blockchain to protect personal data. In: *2015 IEEE Security and Privacy Workshops*. New York, NY: IEEE; 2015. p. 180–184.

[10] Chawathe SS. Clustering blockchain data. In: *Clustering Methods for Big Data Analytics.* New York, NY: Springer; 2019. p. 43–72.

[11] Idelberger F, Governatori G, Riveret R, *et al.* Evaluation of logic-based smart contracts for blockchain systems. In: *International Symposium on Rules and Rule Markup Languages for the Semantic Web*. New York, NY: Springer; 2016. p. 167–183.

[12] Roy S, Sarkar D, Hati S, *et al.* Internet of Music Things: an edge computing paradigm for opportunistic crowdsensing. *The Journal of Supercomputing.* 2018;74(11):6069–6101.

[13] Koulu R. Blockchains and online dispute resolution: smart contracts as an alternative to enforcement. *SCRIPTed.* 2016;13:40.

[14] Xu X, Weber I, Staples M, *et al.* A taxonomy of blockchain-based systems for architecture design. In: *2017 IEEE International Conference on Software Architecture (ICSA)*. New York, NY: IEEE; 2017. p. 243–252.

[15] Johnson D, Menezes A, and Vanstone S. The elliptic curve digital signature algorithm (ECDSA). *International Journal of Information Security.* 2001; 1(1):36–63.

[16] Lu Q and Xu X. Adaptable blockchain-based systems: a case study for product traceability. *IEEE Software.* 2017;34(6):21–27.

[17] Zhang Y and Wen J. An IoT electric business model based on the protocol of bitcoin. In: *2015 18th International Conference on Intelligence in Next Generation Networks*. New York, NY: IEEE; 2015. p. 184–191.

[18] Atzori L, Iera A, and Morabito G. The internet of things: a survey. *Computer Networks.* 2010;54(15):2787–2805.

[19] Sundmaeker H, Guillemin P, Friess P, *et al.* Vision and challenges for realising the Internet of Things. Cluster of European research projects on the Internet of Things, *European Commission.* 2010;3(3):34–36.

[20] Dai HN, Zheng Z, and Zhang Y. Blockchain for Internet of Things: a survey. *IEEE Internet of Things Journal.* 2019;6(5):8076–8094.

[21] Bera S, Misra S, and Vasilakos AV. Software-defined networking for internet of things: a survey. *IEEE Internet of Things Journal.* 2017; 4(6):1994–2008.

[22] Sharma PK, Singh S, Jeong YS, *et al.* Distblocknet: a distributed blockchains-based secure SND architecture for IoT networks. *IEEE Communications Magazine.* 2017;55(9):78–85.

[23] Namasudra S, Sharma P, Crespo RG, *et al.* Blockchain-based medical certificate generation and verification for IoT-based healthcare systems. *IEEE Consumer Electronics Magazine.* 2022;12(2):83–93.

[24] Das S and Namasudra S. Multi-authority CP-ABE-based access control model for IoT-enabled healthcare infrastructure. *IEEE Transactions on Industrial Informatics.* 2022;19(1):821–829.

[25] Namasudra S, Deka GC, Johri P, *et al.* The revolution of blockchain: state-of-the-art and research challenges. *Archives of Computational Methods in Engineering.* 2021;28(3):1497–1515.

[26] Agrawal D, Minocha S, Namasudra S, *et al.* A robust drug recall supply chain management system using hyperledger blockchain ecosystem. *Computers in Biology and Medicine.* 2022;140:105100.

[27] De D, Karmakar A, Banerjee PS, *et al.* BCoT: introduction to blockchain-based Internet of Things for Industry 5.0. In: *Blockchain Based Internet of Things.* New York, NY: Springer; 2022. p. 1–22.

[28] Karmakar A, Ghosh P, Banerjee PS, De D. ChainSure: Agent free insurance system using blockchain for healthcare 4.0. *Intelligent Systems with Applications.* 2023;17: 200177.

[29] Bhattacharyya S, Banerjee PS, Karmakar A, *et al.* BCoT: concluding remarks. In: *Blockchain Based Internet of Things.* New York, NY: Springer; 2022. p. 289–293.

[30] Pop C, Cioara T, Antal M, *et al.* Blockchain based decentralized management of demand response programs in smart energy grids. *Sensors.* 2018; 18(1):162.

[31] Ray PP, Dash D, and De D. Real-time event-driven sensor data analytics at the edge-Internet of Things for smart personal healthcare. *The Journal of Supercomputing.* 2020;76(9):6648–6668.

[32] Griggs KN, Ossipova O, Kohlios CP, *et al.* Healthcare blockchain system using smart contracts for secure automated remote patient monitoring. *Journal of Medical Systems.* 2018;42(7):1–7.

[33] Yang Z, Yang K, Lei L, *et al.* Blockchain-based decentralized trust management in vehicular networks. *IEEE Internet of Things Journal.* 2018;6 (2):1495–1505.

[34] Kang J, Yu R, Huang X, *et al.* Blockchain for secure and efficient data sharing in vehicular edge computing and networks. *IEEE Internet of Things Journal.* 2018;6(3):4660–4670.

[35] Shi S, He D, Li L, *et al.* Applications of blockchain in ensuring the security and privacy of electronic health record systems: a survey. *Computers & Security*. 2020;97:101966.

[36] Tanwar S, Parekh K, and Evans R. Blockchain-based electronic healthcare record system for healthcare 4.0 applications. *Journal of Information Security and Applications*. 2020;50:102407. Available from: https://www.sciencedirect.com/science/article/pii/S2214212619306155.

[37] Bodkhe U, Tanwar S, Parekh K, *et al.* Blockchain for Industry 4.0: a comprehensive review. *IEEE Access*. 2020;8:79764–79800.

[38] Jin H, Dai X, Xiao J, *et al.* Cross-cluster federated learning and blockchain for Internet of Medical Things. *IEEE Internet of Things Journal*. 2021;8(21):15776–15784.

[39] Abu-elezz I, Hassan A, Nazeemudeen A, *et al.* The benefits and threats of blockchain technology in healthcare: a scoping review. *International Journal of Medical Informatics*. 2020;142:104246. Available from: https://www.sciencedirect.com/science/article/pii/S1386505620301544.

[40] Wang Y. Dewblock: a blockchain system based on dew computing. *Architecture*. 2018;3:4.

[41] Papadopoulos GT, Antona M, and Stephanidis C. Towards open and expandable cognitive AI architectures for large-scale multi-agent human-robot collaborative learning. *IEEE Access*. 2021;9:73890–73909.

[42] Heo J, Jang JG, Kwon D, *et al.* Quantum identity authentication with single photon. *Quantum Information Processing*. 2017;16(10):1–20.

Chapter 6

IoT and blockchain technology-based healthcare monitoring

Jose Sergio Magdaleno-Palencia[1], Bogart Yail Marquez[1], Angeles Quezada[1] and Arnulfo Alanis[1]

Abstract

Nowadays, there are several Internet of Things (IoT) protocols with different purposes and application areas, such as transportation, logistics, industry, smart homes, robotics, and healthcare. Since there are advantages and disadvantages to each of them, it is necessary to use the best technologies for people's health and personal data protection. This paper will attempt to review the IoT protocols used in the e-Health industry and the topologies used in each of them. This review aims to introduce the existing technologies, e-Health topology, and their advantages. With technological advances, IoT has become much more accessible to the market, offering a branch of products that standardize and facilitate the adoption of these. Most home devices are considered part of IoT, bringing many social facilities and advantages. The correct adoption of IoT can be very beneficial for society. A little beyond commercial, implementing IoT in monitoring devices can save lives. From intelligent navigation systems that can alert drivers of a risk of collision in their vehicle to real-time monitoring of the critically ill patient. There are innumerable benefits of this technology if it is well applied. In the last decade, one of the fastest-growing industries has been medicine.

Keywords: Cryptography; Blockchain; Health monitoring; e-Healthcare; IoT devices

6.1 Introduction

In this technological era, blockchain is one of the advanced technologies. Blockchain is a technology that mainly involves a decentralized network of nodes.

[1]Departamento de Sistemas y Computación, Maestría en Tecnologías de la Información, Tecnológico Nacional de México campus Tijuana, México

A decentralized network is a set of nodes without any central authority or single points of failure in which decisions are made based on a consensus or joint agreement [1]. As the name indicates, blockchain is nothing more than a chain of blocks. The main characteristics of blockchain are immutable, distributed, decentralized, secure, transparent, and many more. However, this novel technology also has numerous issues that need to be solved before applying it to blockchain-based solutions. Some major challenges are scalability, cost, complexity, and energy consumption.

On the other hand, Internet of Things (IoT) deals with physical objects with sensors, processing ability, software, any electronic device, and other technologies that connect and interchange data or file with other systems and devices over the internet. The key features of IoT are connectivity, scalability, artificial intelligence, dynamic nature, security, integration, analyzing, endpoint management, and compact nature of devices. IoT has many real-life applications like creating better enterprise solutions, smart homes, innovating agriculture, building smarter cities, advanced supply chain management, and many more. Nowadays, IoT devices support sending data to private blockchain networks for creating tamper-proof shared transaction records.

In recent years, applications of IoT and Blockchain in the healthcare sector is continuously increasing. Healthcare is growing to support a patient-centered approach. Here, patients always want to have control over their medical data or any kind of healthcare-related information. In this IoT era, data are directly collected from patients by using some IoT devices and sensors for their treatment purposes. After that, data are processed for medical examination, and then, they are sent to doctors or other medical professionals. Concerns arise when the same patient's record is held by multiple vendors in different versions that have not been validated, leading to various mistakes, inconsistencies, and incompleteness. Currently, medical data are transferred over the Internet, and there are numerous attackers over the Internet. Patients' data can be utilized or sold by unauthorized people for earning revenue or some other personal aspects. So, patients' data can be exposed to anyone. Thus, it is now crucial to maintain data security and stop illicit activity. Utilizing blockchain technology can address these challenges and concerns of any modern healthcare framework by ensuring data integrity and preventing manipulation and failure at any one point.

Many researchers are working to improve the data security of any healthcare framework using blockchain and IoT. The transparency and communication between patients and healthcare professionals have improved because of the introduction of blockchain technology. Moreover, blockchain technology supports immutability property. So, no one can tamper data of a blockchain and IoT-based healthcare system. While blockchain has the potential to improve efficiency and create business opportunities in the healthcare sector, it also requires regulation, information security, and transparency [2]. For example, while sharing patients' Electronic Health Records (EHR) helps to speed and improve diagnosis, the privacy and security preservation of such data is imperative. Node scalability, data storage, and social and cultural issues, such as doubts about how the technology works or resistance to structural changes in the healthcare sector, are a few of the challenges faced by blockchain in healthcare. Hence, it is clear that there are scenarios in which it is hard to adopt the technology either due to technical, social, or regulatory issues. To maintain confidentiality and achieve fine-grained access control, Ciphertext Policy Attribute-Based Encryption (CP-ABE) technique is very commonly used in an IoT-based healthcare system [3–5].

The rest of the chapter is divided into seven parts. Section 6.2 discusses many IoT devices that are used to monitor patients' health. Then, Section 6.3 discusses data collection techniques using IoT devices. Here, the impacts of many advanced technologies like artificial intelligence and big data in the healthcare sector are discussed. e-Healthcare IoT topology is discussed in Section 6.4. Sections 6.5 and 6.6 discuss remote patient monitoring and fake drug detection techniques, respectively. Finally, Section 6.7 concludes the entire chapter.

6.2 IoT devices for health monitoring

It is necessary to visualize that the IoT has a starting point that are connected the devices to the Internet. Considering the different factors for their intercommunication, all this could generate an impulse of how these can be a point of growth in marketing in the medical area. It is also necessary to contemplate and analyze many of the growths in devices that can be connected to generate a network of ubiquitous body sensors. RFID technology has been incorporated into a wide range of products from 2008 to date, which must be considered. More and more, the medical industry is at the point of becoming aware of what it would support and will benefit the medical industry [6].

The Internet of medical things is a set of devices (sensors) that can and will generate a significant improvement and treatment of massive amounts of data. These same ones can currently support and help analysis, classification, and prediction, for the support of a massive study of the data to support the follow-up and treatment of diseases that the users generate.

Given the large volumes of information, medicine presents one of the significant challenges covered by artificial intelligence techniques, such as big data, data mining, and even learning. Deep mining supports a more robust IoTM in health systems. With the above, the IoMT supports the health sectors, with the union of hardware and software to support the growth of Digital Transformation in the Sector [7,8].

Telemedicine is responsible for providing medical and health services remotely. In this way, it is possible to provide health care in inaccessible areas, poor or places where there is no technology or qualified doctors in a specific area of medicine, with IoMT patients can be monitored remotely without having to travel to medical centers for control. These devices allow telemedicine to be interactive, allowing patients and doctors to communicate and interact in real-time [9].

Wearables are medical devices that patients can wear to measure their vital data. All this information is sent to the cloud by these IoMT devices. Thus doctors and health professionals have real-time access to them (without needing the patient or doctor to be in a health center or hospital). For example, the effects of an arrhythmia can be severe, but they are avoidable in most cases. To anticipate these problems, a textile garment with built-in electrodes has been developed. It is a less invasive formula for patients since it does not require cables or adhesives [10].

This work explains the implementation of an IoT-based In-hospital healthcare system using ZigBee mesh protocol. The healthcare system implementation can periodically monitor the physiological parameters of In-hospital patients [11]. It discusses monitoring patients' temperature, respiration rate, heartbeat, and body movement using a Raspberry Pi board. Information output was obtained using a monitor connected to the device on which patient information is displayed [12]. Also introduces the review of the Internet-based Healthcare Monitoring System (HCMS) and the general outlines of opportunities and challenges of the patient's Internet-based patient health monitoring system [13].

A mobile device-based wireless healthcare monitoring system was developed, which can provide real-time online information about the physiological conditions of a patient. The patient's temperature, heart beat rate, and EEG data are monitored, displayed, and stored by the system and sent to the doctor's mobile containing the application [14].

6.3 e-Healthcare data collection using IoT devices

The Internet of Things, or IoT, aims to connect everyday devices to the Internet network and, in this way, obtain through sensors. This information can be processed and analyzed for different purposes, for example, monitoring, location, computing in the cloud, and Artificial Intelligence (AI), among others. Kevin Ashton first used the term in 1999, and later spread worldwide by the Auto-ID Center, a research group working in radio wave identification and other emerging technologies [15]. Some of the main challenges of this technology are:

- New business models, IoT-based currencies, and trust.
- Ethics, control of society, surveillance, and consent.
- Technological challenges, such as energy use, security and privacy, communication mechanisms, and integration of intelligent components, in non-standard substrates
- It is finding a balance between planning and innovation.
- In this context, it can currently find practical applications for health monitoring; taking advantage of IoT, information is obtained from people to be analyzed later, as seen below.

6.3.1 Applications of AI and big data in e-healthcare data collection

Big data is the ability to collect, process, and interpret massive amounts of information. From the health sector, the question is how to access, distribute and use this vast amount of unstructured data. Data that patients, doctors, and health centers have on paper or in electronic format but that are not used due to the impossibility of accessing them effectively. Patients' names, ages, sexes, etc. are the structured data. In contrast, the unstructured data are their medical records, handwritten notes, voice recordings, X-rays, CT scans, and other medical images. Various devices, sensors, medical devices, hospital data, and genetic and genomic information are generating enormous amounts of data due to technological advances.

AI is a resource that allows you to reduce resources and process large amounts of data, and in some cases, make better decisions. In simple terms, AI refers to systems or machines that mimic human intelligence to perform tasks and can iteratively improve the information they collect. Many companies already use this technology to help them know what the consumer prefers. For example, when we buy online, the same application does not suggest items previously searched. Or accessories for the item to buy, chatbots that help understand a customer's problem and provide more assertive responses.

Artificial intelligence can be used not only to be applied to commerce but also in medicine. Accurately predicting the need for critical care is vital for the early identification of vulnerable and high-risk patients [4]. If they have a deep knowledge of the symptoms that patients have would help us to make a diagnosis and thus speed up the treatment of the patient.

Artificial intelligence in telemetry: what clinicians need to know? A patient's situation is described in detail, mentioning the temperature, electrocardiography, pulse oximetry, blood pressure, electroencephalography, and all the studies that must be done. It is detailed that there are a large number of diagnostic tools that can be used to collect information. AI is already an important element in medical care, for example, in the evaluation of skin lesions, the evaluation of fundus retinography for the detection of diabetic retinopathy, and radiological diagnosis, for example, the interpretation of chest X-rays. When patients go to a medical center or are taken by an ambulance, they are not always well evaluated, either due to inexperience of the staff, lack of staff, or poor classification in triage. Triage quickly screens arriving or emergency patients and classifies them. The system divides the severity states, including from critical to less urgent situations. The most urgent patients are seen first, and the rest are reevaluated until the doctor can see them.

Patients classified as red need immediate treatment, while blue patients are neither urgent nor admitted to the hospital. They are evaluated with vital signs, such as temperature, capillary blood glucose, eye-opening, verbal response, motor response, essential neurological functions, radial pulse status, respiratory rate and depth, drowsiness, or confusion. With abundant data available, Data Science is revolutionizing the health sector. Discover how data analytics and AI are transforming the medical industry and how to become a Healthcare Data Scientist.

The health sector generates immense amounts of data. According to a study by the Ponemon Institute, this sector alone represents 30% of global data. Medical records, clinical tests, genetic information, bills, connected objects, databases, and scientific articles, are just a few examples of countless data sources available to the medical sector. With the rise of remote consultations and Internet searches on health topics, the data volume has exploded. For industry professionals, patient data is now centralized and more accessible than ever.

Currently, "Quantified Health" integrates data from connected objects, such as bracelets, watches, and accessories, such as glucometers and scales, in medical records through smartphones. Connected objects are what platforms like Apple HealthKit and Google Fit propose. With these resources, it is possible to quickly detect alarming symptoms and more closely monitor changes in behaviors and vital signs. Health professionals can exploit all these data, opening up a universe of possibilities. Discover how Data Science revolutionizes the health sector.

On average, it takes between 2.6 billion dollars and 12 years to create and bring a drug to market. However, Data Science makes it possible to reduce the cost or the time needed drastically. The data scientists can simulate the reaction of a drug with proteins in the body and different types of cells. According to Mark Ramsey, Chief Data Officer of the pharmaceutical giant GSK, the process can be reduced to less than 2 years thus because of this simulation method. Several start-ups are also exploring this idea. For example, London-based BenevolentAI has raised $115 million to launch more than 20 drug creation programs and develop an artificial brain capable of creating new drugs and treatments.

The connection of objects and other monitoring devices, which take into account the history and genetic information of the patient, can detect problems before they are irreversible. The company Omada Health uses, for example, connected accessories to create personalized behavior plans and online coaching. This help prevents chronic diseases, such as diabetes, hypertension, and cholesterol.

For its part, the Canadian startup Awake Labs collects data from autistic children through connected accessories. With IoT e-Health care, parents can be alerted in case of crisis risk. Nowadays, unfortunately, medical diagnoses are still sometimes wrong. According to the National Academy of Sciences, Engineering, and Medicine, approximately 12 million Americans are misdiagnosed. The consequences can sometimes be fatal. According to a survey by the BBC, diagnostic errors cause between 40,000 and 80,000 deaths annually. However, Data Science makes it possible to improve the accuracy of these diagnoses significantly. In particular in the case of medical imaging analysis. Computers can learn to interpret MRIs, X-rays, mammograms, and other types of rays. The machine learns to identify patterns in the visual data and can then detect tumors, arterial stenosis, and other anomalies with an accuracy that surpasses that of human experts.

Even without reaching the point of automated image analysis, Data Science allows images to be enlarged and sharpened, making image interpretation easier for human experts. On the other hand, researchers at Stanford University have developed Data-Driven models to detect heart rhythm irregularities from electrocardiograms faster than a cardiologist. Other models are capable of distinguishing between benign marks on the skin and malignant lesions.

For their part, Microsoft researchers analyzed web search data from 6.4 million Bing users whose search results suggested they had pancreatic cancer. It is possible to propose more specific and personalized treatments in Data Science. It is possible to take into account subtle differences between each of us to achieve more effective treatments. For example, the 1,000 Genomes project of the National Institute of Health is an open study of regions of the genome associated with common diseases, such as diabetes or coronary heart disease. This study allows scientists to understand better human genes' complexity and how a specific treatment will best suit an individual. For its part, SeamlessMD develops a platform for post-operative care. This platform enables Healthcare System in Saint Peter in, New Jersey, to reduce the average post-op patient stay to one day. This average represents a saving of 1,500 dollars for each patient. Those who only need to indicate their pain level in the application every day and let the doctors monitor their evolution over time. In case of a potential problem, the application issues an alert. Mobile applications that use AI can also help patients. Chatbots or virtual voice assistants can communicate with patients, describing their symptoms or asking questions and receiving valuable information drawn from a vast network that connects signs with diseases.

These applications can also remind the patient to take their medicine at the indicated time and schedule an appointment with a doctor, among other things. Among the most popular are the Woebot chatbots developed by Stanford University to help depressive patients or the virtual assistant of the Berlin startup Ada that predicts diseases from symptoms. Hospitals are establishments in which management is complex and challenging. Data analysis makes it possible to determine how many doctors and nurses must be present daily and at each moment to ensure effectiveness.

Data Science also makes it possible to ensure that enough beds are available to meet demand and much more. Predictive analysis also makes it possible to optimize plans and make the emergency service more fluid. At Emory University Hospital, Data Science predicts the demand for laboratory tests. This data science reduces the waiting time by up to 75%. Business Intelligence can also be exploited to improve the billing system and identify patients at risk of having difficulty paying. These analyses can be coordinated with insurers and financial departments. In this way, the Center for Medicare and Medicaid Services has saved 210.7 million dollars, thus with the prevention of fraud based on Big Data.

The health industry is in complete transformation thus because of data science. Pharmaceutical giants, biotech startups, research centers, and health establishments are investing more and more in this revolution. Many challenges remain to be addressed. For example, data is often scattered across multiple regions, administrative units, and hospitals. This data makes it difficult to consolidate into a single system.

Additionally, many patients are concerned about the protection and privacy of their data. Some private companies are interested in exploiting this valuable data for advertising targeting. Google, in particular, has been sued for these practices. Finally, it is feared that the relationship between doctors and patients will disappear in favor of interactions with machines and algorithms. Human contact is indeed essential in the field of health. Be that as it may, despite the difficulties to be overcome, Data Science offers many promises for the future of medicine. As technology develops, new possibilities will appear. The role of a health data scientist is to design studies and evaluations, perform complex data analysis, or advise healthcare institutions and caregivers based on the results of their research.

It is necessary to use the data to predict the effects of drugs and understand the diseases that affect humans. Its role is also to deploy the power of artificial intelligence and enrich public health data sets. This professional may work for government health departments, hospitals, universities, research institutes, pharmaceutical companies, health insurance companies, or private companies. Becoming a health data scientist requires the same skills as a regular data scientist. However, these skills must be accompanied by a solid knowledge of the field of health.

6.3.2 e-Healthcare data collection using IoT devices

IoT in early detection of Covid-19 based on respiratory rate and body temperature in the last 2 years; it has been seen how the world has been involved in a pandemic caused by Covid-19. Humankind has witnessed how dangerous it can be to contract the virus. Therefore, the importance of early detection, however, diagnosis can become late, especially in places where there is no easy access to tests or there is a saturation of health centers; which means that people do not receive necessary care on time, thus harming their health and putting their lives at risk. This study aims to detect viruses early through IoT sensors of body temperature and respiratory rate; the study applies an exploratory research methodology to analyze a problem in pre-diagnosis. This is carried out through a previous investigation of database data where patterns of vital signs are identified, and this pre-diagnosis is done remotely. Hence, it is not necessary for a patient to perform other types of tests, also avoiding contact [16].

The use of cloud and IoT in combination for healthcare-based systems provides better structure and utilization of monitoring systems. This combination is done through interconnected applications that exchange information to deliver efficient clinical solutions. Sensor-based data collection has grown too large in recent years, complicating data processing and storage. To solve these adversities, the fusion of these two technological tools allows us to provide an efficient solution for device communication and increase the demand for data in health applications. The IoT Cloud framework enables us to develop applications seamlessly and deliver services using cloud-based models. Such aspects usually address data collection, transmission, processing, and storage. This cloud framework helps identify actors and data flow responsible for transforming data provided for real-time streaming to the cloud [17]. Only some things are positive because these services often have many vulnerabilities that affect IoT infrastructure and vulnerabilities inherent in the cloud. These could render health care services non-functional and critical to patient information that could be misused. However, various ways of resolving these insecurities are known, so these platforms are becoming increasingly viable.

Before IoT, many things were limited; one of them was medicine. Interaction between patients and doctors was limited to just visits and communications by text or phone. There was no possible way that doctors could monitor the health of their patients, so there was no way that recommendations could be made according to the current state and time of the patient. This is where IoT comes in, which has made many things possible, including devices that have allowed remote monitoring throughout the healthcare sector, unleashing the full medical potential for doctors to keep patients safe and healthy and to be able to provide better care. IoT has undoubtedly transformed the healthcare industry in a way that can be used in many ways [18]. These IoT applications benefit patients and families, hospitals, doctors, psychics, insurance companies, etc. For patients, numerous devices give patients access to personalized care. These devices can be programmed for various functions, such as blood pressure checks, exercise checks, calorie counting, appointments, and many other things that are helpful to patients and make their care more accessible.

In the case of doctors, with IoT devices, they can monitor and maintain a constant review of patients' health more effectively. They can check if patients continue with the necessary treatment and if there is an improvement, in addition to also being able to check if they need urgent medical attention. In addition to monitoring the health of patients, it also helps hospitals and health centers since IoT devices usually have sensors that are used to track the location of these medical devices in real-time, such as defibrillators, wheelchairs, and oxygen pumps, among many other medical monitoring devices. In addition to being able to count on hygiene monitoring IoT devices to keep patients safe from contracting an external disease or even contracting an infection [19]. All these are benefits that IoT has brought to healthcare, which has come to improve everything for patients, doctors, hospitals and health centers, etc.

Artificial Intelligence and Big Data

IoMTs generate a large amount of information about patients transmitted via the Internet, communicating between devices, computer systems, or healthcare professionals themselves. This communication generates an immense amount of data (i.e., big data) that can be processed and organized intelligently using artificial intelligence algorithms. To convert all this information into knowledge that generates value for health. Some examples of benefits provided by the intelligent analysis of all the data generated by IoMT devices are found in the early detection of diseases as a result of detecting their symptoms at the right time, making diagnoses and providing adequate treatment, or identifying risk factors for suicide, alcohol abuse, or drug abuse (preventive medicine) [20]. Speaking of something more recent like Covid-19, 80% of patients with this virus require outpatient hospitalization since they are mild cases. The problem is that because hospital facilities are overcrowded, it is necessary to consider other measures to prevent more patients from dying from this disease. One of those measures designed is a hardware-based monitoring system with IoT with the ability to obtain parameters, such as respiratory rate and heart rate, among other things [21,22].

Another application is Smart-Steering, an IoMT device for monitoring blood alcohol concentration using physiological signals. Highway accidents are severe and unavoidable incidents that have exponentially increased yearly in the United States. Statistics show that among all accidents, 40% are due to driving drunk or under the influence of alcoholic beverages. With the growth of science and technology, robust solutions with improved, reasonable, and feasible mechanisms must be proposed. Smart-Steering strives to solve this problem with a scope of reduction in the increasing rate of traffic accidents [23].

The application proposes a smart "thing" that can convert a regular address into a smart address with the help of the Internet of Things. This device works with the touch of the human driver, collects, and analyzes physiological data of the person, and performs analysis on a microcontroller. Analyzed data, a human's sobriety decision is made, and car information is sent as a notification; this data is also sent to the cloud server for storage purposes. Prediction of alcohol in the blood is made with an exact level of the concentration present in the human body with an accuracy of approximately 93% [23].

Data security has become increasingly significant as data are shared through Cloud in heterogeneous devices [24]. Cryptography is a technique of secret writing [25]. The main goal of cryptography is to transmit data or message between sender and receiver over an untrusted medium so that an attacker or malicious user cannot read the original data content. It can also be used for the authentication of customers and users. In Cryptography, plaintext refers to ordinary readable data or messages [26]. The encryption process takes plaintext as input and generates ciphertext as output using a secret key and Encryption algorithm. The process of recovering the plaintext from ciphertext using the secret key is known as decryption.

IoT devices used for health monitoring can improve efficiency in monitoring treatment, as well as recurrent monitoring of patients for early detection of some abnormalities in their health. Thirty percent of medical patients are readmitted to hospitals after surgery. The question every doctor ask is, "why?" The direct answer is the lack of patient follow-up (EGOS BI). Remote patient monitoring is one of the most widely used applications for healthcare IoT devices. Continuous collection of patient data related to health is collected automatically, allowing the device to store data, such as blood pressure, glucose level, temperature, etc., without requiring patient interaction [27]. When necessary, data is collected and sent to an application that uses an algorithm to determine treatment or generate an early warning regarding a patient's health. One of the most critical points to consider regarding the transfer of patient data is their privacy; it must be ensured that the transfer is carried out securely and privately.

An example of a device for health monitoring using IoT is a temperature monitor and heart attack detection, developed with the help of Arduino. Arduino is an element of easy connectivity to a network and additionally allows the implementation of a high-level protocol server, such as the Hypertext Transfer Protocol (HTTP). It has memory, autonomous processing power, compilers for programming languages, such as C, and physical ports to interface with devices [28]. A prototype was developed using an Arduino board and pressure and pulse sensors as a base. Every two milliseconds, Arduino reads an analog signal from a sensor, which is converted into a digital signal. Healthcare applications can translate and interpret this signal to determine a patient's health with constant monitoring. Nowadays is the internet age, and basically, any object can be connected to another to send information. Health care, personal training, and lifestyle are some of the points where health care is complemented using IoT.

e-Healthcare data

The healthcare sector has been strongly affected by IoT. Medical parameters and vital functions can be monitored in real-time by utilizing devices with advanced detection capabilities. This collected data is transmitted through communication technologies (e.g., Bluetooth, ZigBee, and Wireless HART) and trained medical personnel for diagnosis and monitoring of patient health. Body Area Networks (BAN), made up of portable devices connected, allow doctors to continue monitoring patients remotely outside the hospital. Other relevant applications are related to identifying materials and medical instrumentation. For example, the application of smart tags to ensure that tracking objects are accurate to prevent equipment from being lost, stolen, or materials from being left inside patients during an operation. The use of these labels is also essential to facilitate the inventory of medical equipment [29].

Another application of IoT in the health field is assisting elderly or disabled people who live alone by detecting falls to them. Sanitary refrigerators control the temperature at which medicines, vaccines, and organs must be kept inside them. Monitoring athletes on their vital signs when they are in high performance or playing on the field helps to know if they can maintain a suitable rhythm for sport. Lastly, ultraviolet ray measurement warns people of overexposure to their bodies, thereby limiting them from continuing to sunbathe for large amounts of time. As technology advances, IoT has become much more accessible to the market, offering a branch of products that standardize and facilitate their adoption. Most home devices can already be considered part of IoT, and this has brought many facilities and advantages to society. The adoption of IoT, used correctly, can be very beneficial for society. A little beyond commercial, implementing IoT in monitoring devices can even save a life. From smart navigation systems that can alert a driver of a risk of collision in their vehicle to real-time monitoring of a critically ill patient [30].

There are innumerable benefits of this technology if it is well applied. In the last decade, one of the fastest-growing industries has been medicine. It is not necessary to look for statistics to prove it, just as wars have been one of the drivers of technological advances in transport and weapons. The start of the Covid-19 pandemic marks a before and after in basic needs Worldwide. And how should it not be that a doctor caring for more than 50 patients in intensive care could improve the quality when the doctor checks the patients. Suppose those 50 people have an intelligent monitoring system that alerts them to a severe change in their condition. A wide variety of medical devices integrate IoT into their function. If analyzed in depth, the principle of medicine is the analysis of data provided by patients or obtained through medical diagnostic devices (thermometer, glucometer, etc.). All this information is collected by the doctor, with whom he makes a diagnosis and prescribes a treatment. With the implementation of IoT in medical devices, the healthcare field is taking it one step further regarding data collection and analysis.

Internet of Things

The Internet of Things or IoT is a system of computing devices, mechanical and digital machines, objects, animals, or people given unique identifiers and can transfer data over a network without requiring human–computer or computer–computer interaction. When referring to Things (from the IoT term), it can be anything from a person with a medical implant to a car with sensors, any device that can be assigned an Internet Protocol (IP) address and can transfer data [31]. An IoT ecosystem consists of a network of web-enabled smart devices that subtract, collect, and use data they collect within their environment. These data must have the ability to be sent to a local or remote database or even interact with other devices within the same ecosystem to improve the quality and accuracy of the data they provide.

6.4 e-Healthcare IoT topology

On a large scale, IoT consists of millions of connected devices and sensors that continuously transmit information. This premise is applied in the health field since it can keep patients connected with their health providers even when they leave the hospital. There are many concepts about how IoT architecture can be divided, and the easiest to identify stages of its architecture are the following [32].

1. *Collection*: Interconnected devices are responsible for collecting information.
2. *Conversion*: Data is received in analog form is transformed into digital data.
3. *Pre-processing*: Data is standardized and it is sent to a data center or cloud.
4. *Analysis*: The final stage is where data is processed and analyzed for its final purpose.

Figure 6.1 shows an IoT-based general architecture.

Considering the architecture described above, the second and third stages could be considered critical foundations of an IoT system. Without the possibility of sharing data obtained, the entire purpose of these devices would not exist. Any IoT system needs sensors to be able to carry out the collection; two key elements are required for its function: the Internet/network and sensors. The sensors are the devices in charge of collecting all the necessary information; the application sets guidelines for the type of sensor that will be used. The most common example of an IoT system is a smartphone that integrates sensors and communication into a single device.

To carry out this data transmission, it is necessary to have a gateway device. This device is responsible for transforming collected data into a communication

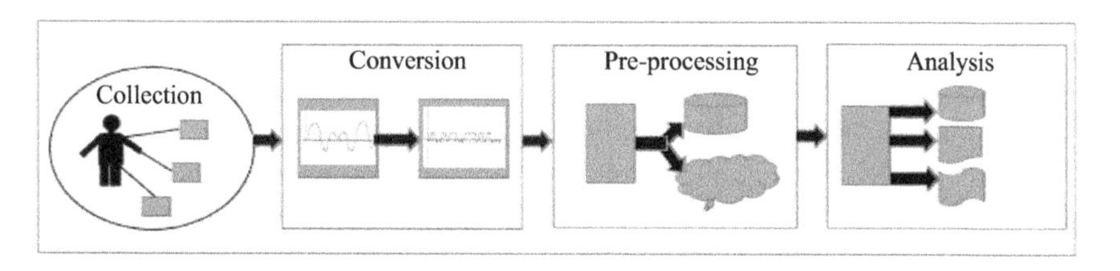

Figure 6.1 IoT general architecture

protocol that can transmit that data to its storage and processing destination. There are several communication methods/protocols used in IoT: Radio Frequency Identification (RFID), Electronic Product Code (EPC), Wireless Sensor Network (WSN), Near Field Communication (NFC), Zigbee, Z-Wave, and Bluetooth LE. A gateway device is responsible for "translating" information collected by IoT devices and sending it to servers/clouds [29]. An emerging concept in IoT for the medical industry is Wearable Sensors. The basic principle of this type of sensor is that anyone can use them and collect information in large quantities. This role goes hand in hand with "Big Data," which refers to considerable and variable amounts in a fast-moving environment that need to be collected, processed, analyzed, and integrated.

Specifically, in the medical area, the main objective of sensors is the monitoring of all health parameters of end users, such as heart rate and its variations, respiratory rate, glucose levels, and blood pressure. The smartwatch is an example of a device that meets most of these sensors. These sensors serve as the front end of an IoT system. They allow health professionals to monitor their patients in real-time or give users a premise about their health status. However, experts argue that this type of wearable sensor can be considered invasive and prefer embedded sensors. In the United States, the IoT market in the e-Healthcare industry has increased in recent years. Figure 6.2 shows IoT's importance in the market. It shows annual growth and forecast according to different fields in the applicable e-Healthcare industry, with the highest percentage of growth being electronic prescription systems.

The Internet of Things is an emerging paradigm that enables the communication between electronic devices and sensors through the internet to facilitate human lives. IoT uses smart devices and the internet to provide innovative solutions to various challenges and issues related to various business, governmental, and public/private

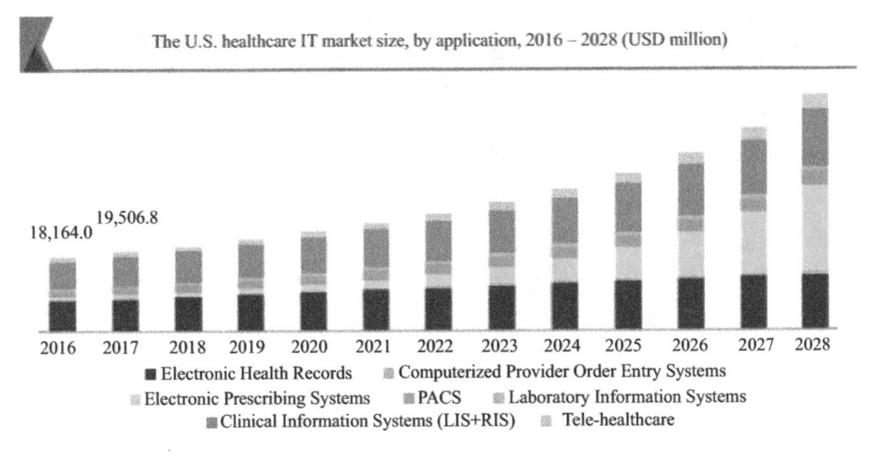

Figure 6.2 Importance of IoT in the market

industries worldwide. In health sciences, these technologies have been developed. In recent years based on their capabilities to monitor various health parameters, which be now transmitted by health devices via a gateway onto secure cloud-based platforms where they are stored and analyzed. The following sections will cover these IoT devices applied to e-Health. What technologies are behind the e-Health platform, what topologies are used by these protocols, where they are being used, and what could be improved.

When health services and information delivered or enhanced through the Internet and related technologies emerge, e-Health, thus with the intersection of medical informatics, public health, and business. This term refers not only to technical development but refers to a state of mind, a way of thinking, an attitude, and a commitment to networked, global thinking, to improve health care locally, regionally, and worldwide by using information and communication technology. RFID systems are used in e-Health, consisting of a reading device and a small radio frequency transponder called an RF tag. Various RFID standards exist, including ISO, IEC, ASTM International, the DASH7 Alliance, and EPC-global. Nowadays, IoT applications using RFID include smart shopping, health care, national security, and agriculture. RFID can support P2P network topology [26].

Figure 6.3 shows that the highly pervasive RFID enables remote identification, tracking, and localization of medical staff, patients, and drugs. And equipment, thus increasing safety, optimizing real-time management, and providing support for new ambient-intelligent services.

Figure 6.4 shows the importance of Z-wave-based devices and systems are centerpieces of the connected aging and wellness paradigm. They allow a new realm of security, convenience, and social features for the elderly that empowers independent and secure aging in place. Utilizing inexpensive sensors, Z-wave can

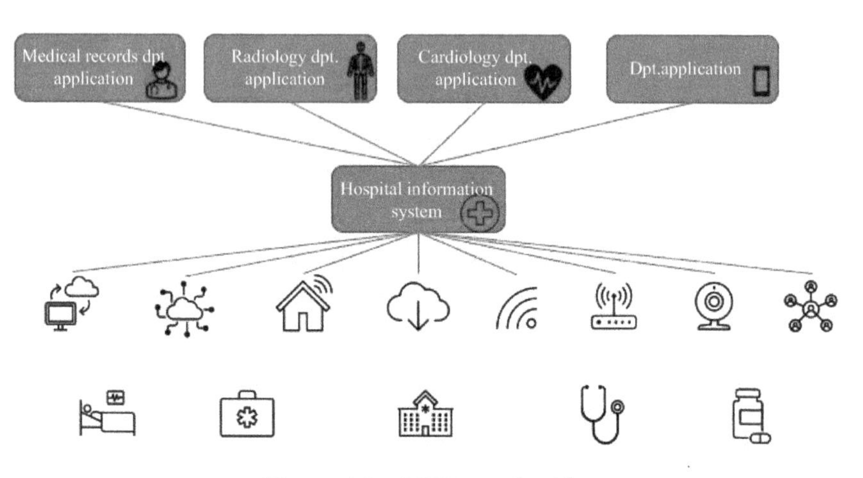

Figure 6.3 RFID in e-health

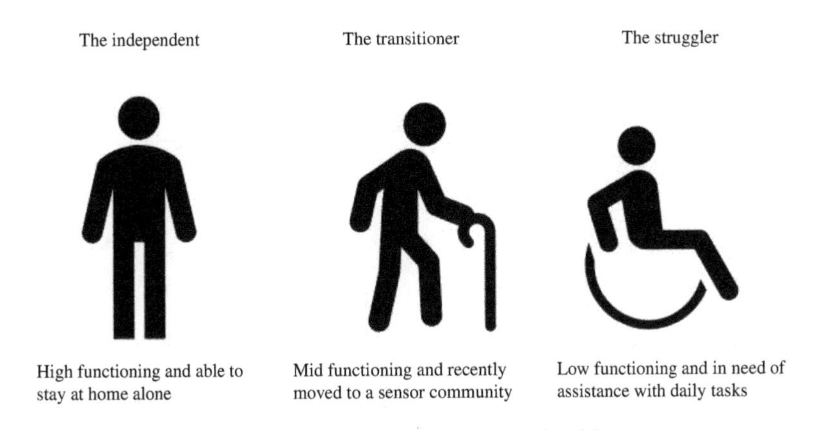

Figure 6.4 Activity is an important new healthcare metric

accumulate and share data about daily wellness activity, ranging from simple ambulation patterns. To how often a medicine chest has been opened or how often the bathroom has been accessed. This allows the remote caretaker to be alerted to any abnormalities in the living patterns of loved ones aging in place [33].

As seen in previous sections, both RFID and Z-wave protocols use a mesh-like topology, P2P and Mesh, respectively. They are similar in how they communicate with the other devices, but there are slight differences. For example, P2P can be configured to work in three different ways centralized, decentralized, partially decentralized, and hybrid. Z-wave works in a Hybrid manner with a Hub managing its nodes.

P2P network topology

The topology of a P2P network refers to the physical or logical interconnection between computing units in a distributed system; topology between nodes has been an essential basis for determining the type of system. This structure allows clients in a network to share some of the computer resources so that the computer can provide its own remaining free resources for other users [34], as shown in Figure 6.5.

Mesh topology is intended to allow a very high level of redundancy by connecting each system within the network to every other system. As more systems are added to the network, the connections between them grow tremendously. This means it is complicated to scale a network using a mesh topology, as shown in Figure 6.6.

Current Z-wave application for e-health

Recently, BDigital developed eKauri, a non-invasive e-health and smart home platform that empowers seniors to gain autonomy, participate in modern

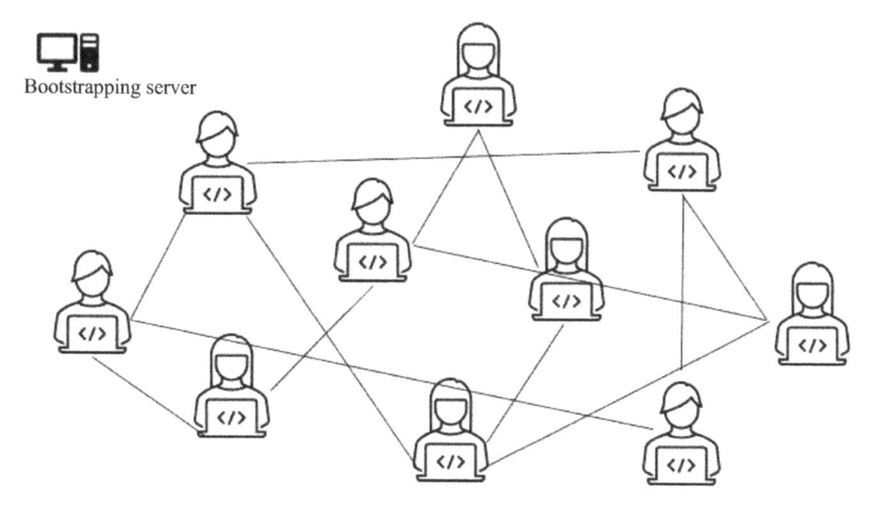

Figure 6.5 Decentralized unstructured peer-to-peer networks

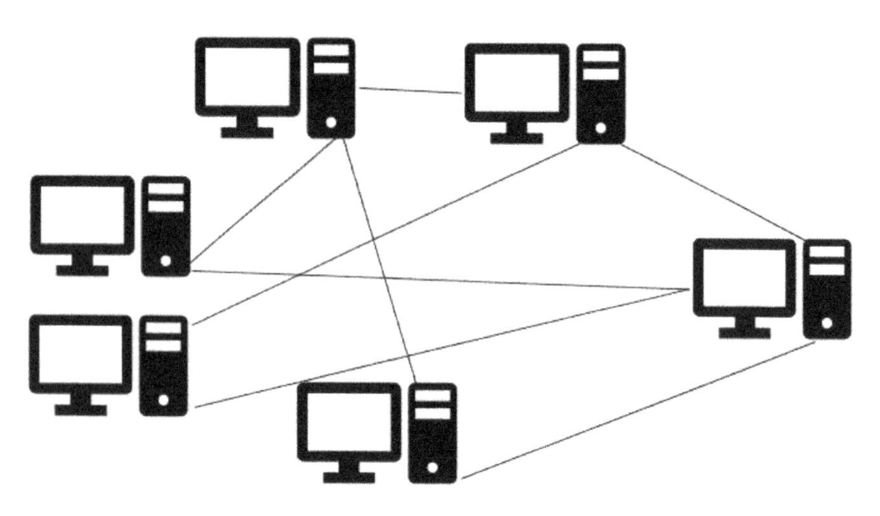

Figure 6.6 Mesh topology

society, and achieve independence through ICT-based solutions. To deliver its services, the cloud-enabled eKauri takes advantage of credit-card-sized Raspberry Pi computers and Z-wave wireless home automation devices. Figure 6.7 shows the eKauri e-health and smart home platform.

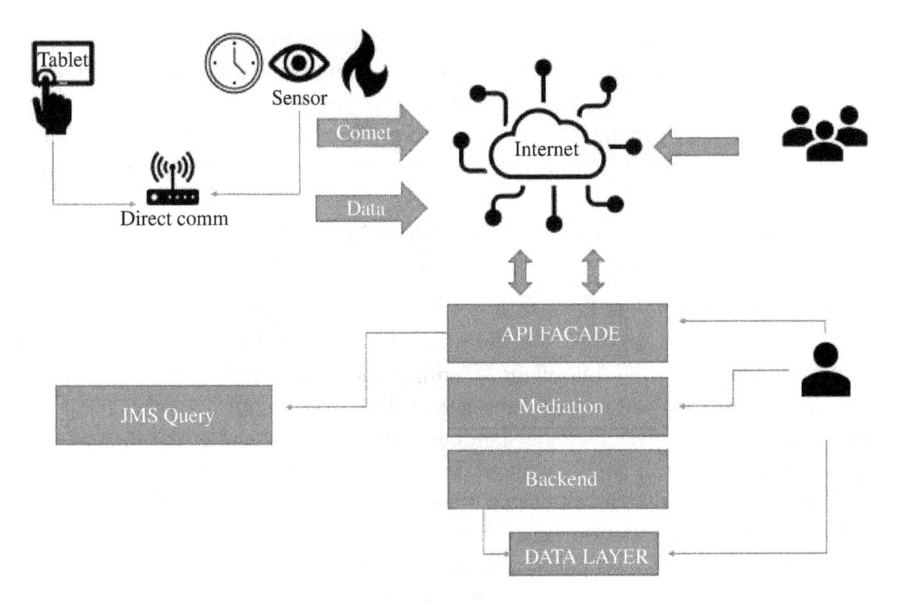

Figure 6.7 The eKauri e-health and smart home platform manage API consumption through an API Façade based on API manager

6.5 IoT and blockchain technology for remote patient monitoring

In 2008, it was born with a technical document in a cryptography forum, in which an electronic money system called cryptocurrency is described, which was based on a distributed and decentralized network that solved the problem of double spending, taking thus the name of bitcoin. Going from a centralized system, where it is necessary to trust a third party for the security of the transactions, to a decentralized one created a significant impact around the world. Originally it was only used in the creation of cryptocurrencies; however, due to its effectiveness, it is being used in other environments where reliable and secure information is required, such as document certification, voting, and other sectors requiring accurate and secure information [35,36].

According to Tapscott [37], prior to blockchain technology, it is necessary to trust third parties, intermediaries, or other entities that serve as third parties. Sometimes, it is not known who these parties were or if they acted with integrity at the time of the process. Blockchain is defined as a distributed and decentralized data registry in which a set of transactions is stored in units called blocks that are linked sequentially over time. All these transactions are verified and approved through a consensus protocol for all members to register them in the blockchain. Once registered, it can no longer be modified without the consent of the majority [38,39]. The main characteristics of a blockchain network are:

1. *Privacy:* allows partial or total anonymity between transactions.
2. *Scalability:* limited by consensus algorithm implemented in the blockchain.
3. *Governance:* agreement from a centralized environment to a decentralized one eliminates risks generated by control and pressure exerted on the system, but it is necessary to modify policies within organizations and adapt methodological frameworks.
4. *Consensus system:* establishes a mechanism by which it tends to guarantee the integrity of information stored in the blockchain.

First, when discussing health care, it is necessary to provide attributes encompassing this concept. for this, the definition by the world health organization is recovered, which says the following: "Health is a state of complete physical, mental and social well-being and not only absence of illness or disease." Another thing that this same document gives us understanding is that health is one of the fundamental rights that every person enjoys without distinctions of any kind. This approach that has been given to health has allowed a change in care model, starting from a doctor-centered model to a patient-oriented model, based on the following six characteristics: a health system with public and interoperable information between different actors, which allows greater patient participation in their treatment and with psycho-cognitive support facilities for the community, focusing health system on preventive, predictive and personalized medicine. Current research is working to implement this model, including care management changes with new stakeholder behaviors.

In this context, blockchain and IoT emerge as the technology enabler that allows the provision of patient-oriented services, with assertive auditing of events and medical care prescriptions. It facilitates the ubiquitous dissemination of medical information with an automated and consistent collection of patient medical data by implementing an authorization mechanism for controlled visualization of clinical data to allow the anonymous exchange of medical information during diagnostic and research procedures. Previously it has been described that health is considered a condition that involves mental, physical, and social well-being. So when it comes to medical care, it is necessary to focus on these three factors from the patient's point of view. This is who lets know the condition better, so knowing how the person feels will allow for better diagnosis and treatment.

The model in question being discussed requires an adequate exchange of medical information between patients, health professionals, research institutes, communities of interest, and family members. Blockchain will allow the design of a secure, reliable, auditable, and ubiquitous scenario to share medical information. Thus, facilitating clinical research processes for a timely and personal patient diagnosis also adds a layer of security to manage medical IoT devices that monitor patients' physical health. In conclusion, an environment with information flow with these peculiarities provides circumstances required for correct comprehensive medical care that allows predictive medicine with the ability to predict the appearance of diseases, predicting possible health states of patients [40]. These conditions can allow access to preventive and personalized medications and tools for early diagnosis, accompanied by therapies adapted to the patient, to achieve greater effectiveness and reduce possible unfavorable effects [38].

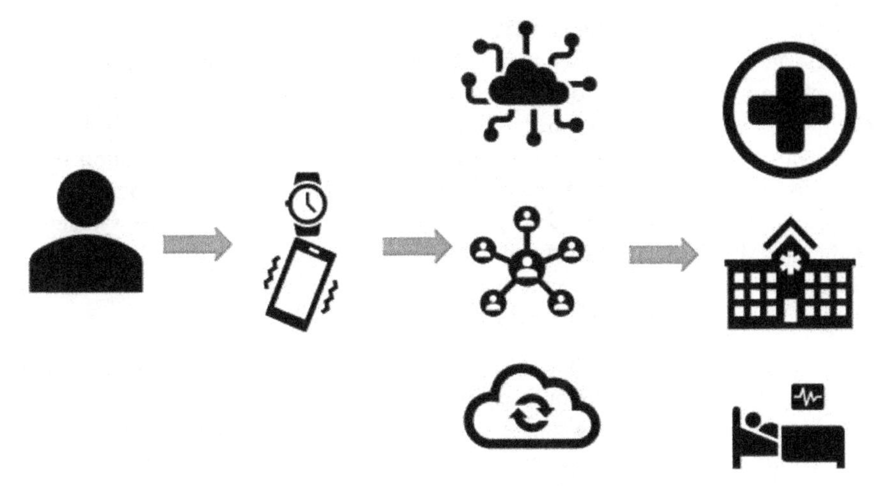

Figure 6.8 IoT devices and blockchain storage

Healthcare IoT (HIoT or e-Health) is experiencing exponential growth in research and industry. However, it still suffers from privacy and security vulnerabilities [41]. Blockchain is distributed immutable ledger without a central authority that has been used recently to provide security and privacy in P2P networks with similar topologies to e-Health. An Interplanetary File System (IPFS) is a protocol designed to store decentralized and shared files enabling a P2P distributed file system to make the web safer, faster, and more open. A peer also stores the patient's vital information in IPFS. That storing results in the IPFS creating a content-based hash that can be decrypted to view the patient's vitality; decentralization will help in the scalability and fault tolerance of IoT devices and data. Decentralization will also prevent certain authority organizations from controlling the flow of information. Figure 6.8 shows the IoT devices and Blockchain storage.

Blockchain defines the human future; as a result of this technology, the Internet will continue to evolve into what is becoming known as the Internet of Value. In recent years, the health sector's demand for equipment and security has increased. For this reason, it was sought how to innovate in this aspect, and blockchain was found. This is made possible by a wireless connection between sensors and devices. Several solutions have been developed due to this distributed registration technology, including document registration and certification, identification of individuals, value registration, and electronic voting. Thus why, together with IoT, they are emerging as fundamental technological support for patient-centered health care.

6.5.1 Blockchain in health sector

As previously mentioned, blockchain allows trust to be generated in a distributed system since transactions, that is, the exchange of information can be carried out directly between interested parties without the need for intermediaries. Due to this, blockchain technology has generated interest in various fields, such as academia and business, with the financial sector being one of the most active. This increased interest opens the door to new models and possibilities so that the health sector can benefit from the potential of blockchain, having multiple use cases. Within the health sector, blockchain has been applied in two main areas, the first of which is the electronic record of medical records, and the second is the verification of the logistics chain of the pharmaceutical industry. Current medical information management systems are expected to be combined with blockchain-type systems to enable the benefits of distributed ledger technology and information security compliance to be deployed. Features that implementing blockchain in an electronic medical record management system will change [38]:

1. *Secure Access to Data:* There are processes to identify and validate patients in blockchain, allowing them to enter anonymously. This adds an extra level of security, thus generating greater confidence when interacting with other users.
2. *Exchange of Medical Information:* Patients will have a unique medical history, making interacting with doctors, specialists, therapists, and pharmacies easier. This history will be stored in centralized databases and requires interoperability mechanisms between blockchain and the health sector.
3. *Government Management of Medical Records:* In the blockchain, citizens, health providers, and companies in the same sector interact, keeping track of medical consultations and records.

Blockchain goes beyond just a distributed ledger; for this, it is necessary to define what information will be stored in the blockchain and then establish rules that this data must comply with. Smart Contracts are used to describe these rules. A Smart Contract is a program created by a user of blockchain that is executed within it to capture agreements between different network users, allowing the management of digital assets. In addition, these contracts enable the automation of transactions that comply with a specific business logic. These transactions are registered in the blockchain so that participants can verify the validity of the information. The three main characteristics of a Smart Contract are neutrality, atomicity, and transparency.

Based on Ethereum code, a blockchain was designed that allows registration and validation of patients, doctors, and health centers; only modifications were made, such as adding a translation service and modifying c the calculation of the public key. Three contracts then govern access and flow of information. First is the identity registration contract, which speaks above all association between the user's identification string and Ethereum address, giving rise to a unique registration. The second contract refers to the interaction between the patient and the provider. Finally, the third contract talks about medical history and withdrawal/re-entry to a user's system. The system modules are described [38]:

1. *Backend Library:* It consists of libraries that abstract communication with blockchain and interact with Ethereum clients.
2. *Ethereum Client:* It is the implementation of a node with the addition of a location to execute contracts.

3. *Database Driver:* It is an interface to communicate with local information repositories of health providers.
4. *Administrator:* The administrator of the electronic medical records system manages the medical records of each provider.

Currently, it has various monitoring devices for the health field; through these devices, a reading of a patient's medical information is obtained, but for this, reliable means are required to ensure a suitable registration and administration of data in the clinical history. All this work has been done with a wireless connection between sensors and devices. Now, in this field of IoT, blockchain can be used to manage smart contracts and, in this way, identify the smuggling or counterfeiting of devices. Therefore, it will be possible to determine when some information from the IoT does not match or is modified in some way. A problem that affects us today is adulteration or counterfeiting of medicines, which has serious repercussions, such as causing a drop in the patient's health and even death, which also causes a loss of confidence in the product or laboratory. A false or falsified drug is one whose ingredients lack active ingredients or do not contain the appropriate proportions of these and can cause serious harm to patients. The measure that can be taken in these cases to prevent the distribution of false medicines is to trace the entire process that has to do with the distribution of the product; in this way authenticity and legitimacy of its origin would be verified, which is known as drug serialization. With blockchain, a shared registry can be created for all interested parties where the life cycle of ingredients and medicines would be registered. In this way, both the distributor and the client can electronically verify the status of the medication or product being tracked, thus checking its authenticity.

Regarding IoT, temperature measurement devices can be used connected to a database and an intelligent device, such as a computer or a cell phone, to record temperature and thus know when a product is not being transported. Under the right conditions, a smart contract would be generated with blockchain, and the product would be invalidated when values outside the allowed range are recorded. The objective of blockchain with the use of IoT in the hospital field is to create a decentralized network, providing security, confidentiality, integrity, availability, and authentication of data obtained. Blockchain seeks to eradicate risks of adulteration of devices or unauthorized access; it is also possible to keep a record of all actions carried out on the IoT. In general, a complete and transparent total control of all information thrown and accessed from devices that make up the IoT network. On the other hand, and returning to previous paragraphs, blockchain can also keep track of the record of the transport of medicines, thus identifying counterfeiting, adulteration, or transportation of products in inadequate conditions [38].

As described above, blockchain technology has come to change the world; it has grown primarily due to its many benefits, one of which is the secure exchange of information without needing a mediator. This has allowed many transactions to be carried out without the help of third parties, which has allowed us to distance ourselves from depending on someone. This has received the name of decentralized transactions, something that later became relevant in many more fields apart from finance, thus being little by little more studied by more organizations with different purposes.

Nowadays, health care has changed its care model from one centered on the doctor to one directed at the patient, so when using this new model, a new series of characteristics has had to be worked on. Under this new framework is that blockchain and IoT have emerged as the major technological enabler that offers patient-driven service delivery with accurate assessments of medical events and prescriptions. This has enabled a reliable flow of personal medical information through automated and consistent data collection through an authorized data viewing mechanism to allow an anonymous exchange of medical data throughout the diagnostic and clinical process research.

The model being talked about requires an adequate exchange of clinical information between patients, health professionals, research institutes, communities of interest, and family members, for which the use of this technology is crucial since it allows a safe, reliable, and verifiable to share medical information. An important factor within this is the use of Smart Contracts that add an extra level of security because rules established in said contracts must be complied with to carry out transactions. Taking advantage of these benefits within the health sector, it finds its primary use case, which consists of electronic management of clinical records, with processes to identify and validate users. That is, patients and health personnel so that interaction between them and the exchange of medical information is allowed. Likewise, patients will have a unique medical history. They will be able to dispose of it regardless of the institution where medical care is performed, which gives patients greater integrity and freedom in making decisions regarding their health care.

6.6 Fake drug detection using IoT and blockchain technology

The growth and the grand opening of online marketing have given rise to the online growth of the traffic of some classified as drugs. Given whatever lived since the beginning of the COVID-19 pandemic, it gave us an increasing breadth of marketing models that could be considered criminal; this does not mean that this did not happen before the pandemic.

An example case is what happens in Latin America and the Caribbean; it is the online trafficking of synthetic drugs and synthetic opioids [4]. A problem that is difficult to eradicate but does not mean that it is impossible is cybercrime or cybercrime, a type of transnational crime that is continually growing and constantly evolving. As it is difficult to measure due to its structure, this allows it to cross borders and for those dedicated to it to have extensive online operations, having a more flexible way of operating, but also in a covert way.

In recent times, there has been a large increase in online drug trafficking, all thanks to the growth of marketing models, given what has been experienced and generated by the Covid-19 pandemic. This boom has gained strength not only on the traditional web but also on the dark web; according to the UNODC World Drug Report 2021, it increased almost four times between 2011 and mid-2017, suffering a more significant increase after that (since mid-2017 to 2020) [5]. The web has allowed many modifications, but has also generated many for marketing and economic activity [9]. All of the above raised in terms of the difficulty of online control, it is necessary to have or contemplate techniques that support this process. Here is where Blockchain technology is being implemented. This technology has incredible growth to generate resolution support in a disruptive way not only to the medical system but also to education and marketing [10]. With them, support the generation and have reliable and reliable online systems.

The blockchain is one of the activities being heard more and more these days due to its impact when validating information. Since, as in the economy, this technology allows for a secure, synchronized, decentralized, and distributed record of operations without the need for a third party. As a result, it has been used in various sectors, including government services, document authentication, voting, and, in this case, monitoring and detecting illicit activities.

The Internet of Things is one of the emerging issues of social, economic, and technical importance since, over the years. It has been combined with consumer products, durable goods, automobiles and buses, industrial and public service components, sensors, and many other everyday objects that can connect to the internet. With IoT, all these devices can communicate and be visible to one another. Virtually anything that a person can imagine could be connected to the internet and interact without human interaction; therefore, it is a machine-to-machine interaction known as an M2M interaction.

Since the first use of the application of blockchain in drug discovery and development, it has managed digital fund transactions. Blockchain technology has found its way into many other industries, with the potential to transform the global economy by changing the way people do business. According to Statista's forecast, international investments in blockchain technology are expected to reach $11.7 billion by 2022. Decentralized information distribution makes blockchain a compelling technology that can transform businesses by providing trust, streamlining processes, reducing operational costs, and eliminating fraud and iterative duplication while enabling data integration in the economy of data. Although the idea has become popular, criminals use encryption technologies (cryptocurrencies) to commit money laundering and cover up purchases of illicit products, such as drugs, among other activities. The information obtained in research shows that less than 1% of these activities use these technologies.

6.6.1 Underused technology

According to a former CIA worker named Morell in a report, it is highlighted that using blockchain to identify illicit activities is a powerful forensic tool. However, the government and authorities are wasting it. It is pointed out that experts in this technology and other authorities in the United States more easily find illicit activities where bitcoin was used than illegal cross-border activities involving traditional transactions. Although new technologies provide tools to hide these activities, most cases occur in the conventional banking system and not in these new technologies. In investigations, it is found that the popularity given to cryptocurrencies for carrying out illicit activities was due to people's ignorance of these technologies for not investigating with reliable sources and the technology that supports blockchain [38].

6.7 Conclusions

IoT is becoming an essential tool of everyday life. It is an opportunity to evolve the care given to patients and the attention given by health professionals to their patients. Functionality and integration of these devices facilitate normalization of use. Since IoT is a technological tool that is constantly evolving, the accuracy of the data it receives and its functionality will increase over time. IoT in e-healthcare is becoming a staple of any device. For it to be intelligent, it is necessary to collect and store data for its processing; over the years, it will give way to better and more efficient solutions. IoT has become one of the technologies that will build the human future.

There are several options for connectivity and protocols, but the industry has been moving to a mesh topology approach due to the advantages of connectivity resilience. However, many improvements are still needed in the e-health area regarding security and personal data encryption, but blockchain will be a viable solution for security and data protection. To finalize as a summary of what has been seen in this document, joint implementation of blockchain and IoT allows a better administration of the clinical data of patients; it also enables monitoring of medicines used by them, thus identifying falsification or transport in inadequate conditions. With blockchain, the patient becomes the total owner of their data, exercising direct control over their clinical history; in this sense, they decide with whom to share their information, all this through smart contracts.

References

[1] M. J. Gámez and O. A. Sánchez, *Proposal for an Innovative Digital Communication System using IoT for Biomedical Patient Monitoring Equipment: Of Interest to the Public Sector*, Santa Tecla, La Libertad, El Salvador: ITCA Publishers, 2021.

[2] A. H. Aman, W. H. Hassan, S. Sameen, and Z. S. Attarbashi, "IoMT amid COVID-19 pandemic: application, architecture, technology, and security," *J. Netw. Comput. Appl.*, vol. 174, p. 102886, 2021.

[3] S. Das and S. Namasudra, "MACPABE: multi authority-based CP-ABE with efficient attribute revocation for IoT-enabled healthcare infrastructure," *Int. J. Netw. Manag.*, 2022. DOI: https://onlinelibrary.wiley.com/doi/abs/10.1002/nem.2200.

[4] S. Das and S. Namasudra, "Multi-authority CP-ABE-based access control model for IoT-enabled healthcare infrastructure," *IEEE Trans. Indust. Inform.*, vol. 19, no. 1, pp. 821–829, 2013. DOI: 10.1109/TII.2022.3167842.

[5] S. Das and S. Namasudra, "A novel hybrid encryption method to secure healthcare data in IoT-enabled healthcare infrastructure," *Comput. Electr. Eng.*, vol. 101, p. 107991, 2022. DOI: https://doi.org/10.1016/j.compeleceng.2022.107991.

[6] D. Miorandi, S. Sicari, F. De Pellegrini, and I. Chlamtac, "Internet of Things: vision, applications and research challenges," *Ad Hoc Netw.*, vol. 10, p. 1509, 2012.

[7] K. Lazarev, "Internet of Things for personal healthcare study of eHealth sector," Smart Wearables Design, 2016. Available at: https://www.theseus.fi/bitstream/handle/10024/119325/thesis_Kirill_Lazarev.pdf.

[8] P. San Martin, K. Avila, C. V. Núñez, and D.J. Molinares, *Internet of Things and Home-Centered Health*, Barranquilla, Colombia: Salud Universidad del Norte, 2016.

[9] M. M. Abdellatif and W. Mohamed, "Telemedicne: an IoT based remote healthcare system," *Int. J. Online Biomed. Eng.*, vol. 16, no. 6, pp. 72–81, 2020.

[10] F. J. Dian, R. Vahidnia, and A. Rahmati, "Wearables and Internet of Things (IoT), applications, opportunities, and challenges: a survey," *IEEE Access*, vol. 8, pp. 69200–69211, 2020.

[11] R. K. Kodali, G. Swamy, and B. Lakshmi, "An implementation of IoT for healthcare," In 2015 *IEEE Recent Advances in Intelligent Computational Systems (RAICS)*, pp. 411–416, 2015.

[12] R. Kumar and M. P. Rajasekaran, "An IoT based patient monitoring system using Raspberry Pi," In *2016 International Conference on Computing Technologies and Intelligent Data Engineering (ICCTIDE'16)*, pp. 1–4, 2016.

[13] K. T. Kadhim, A. M. Alsahlany, A. M. Wadi, *et al.*, "An overview of patient's health status monitoring system based on Internet of Things (IoT)," *Wirless Pers. Commun.*, vol. 114, pp. 2235–2262, 2020.

[14] C. Senthamilarasi, J. J. Rani, B. Vidhya, and H. Aritha, "A smart patient health monitoring system using IoT," *Int. J. Pure Appl. Math.*, vol. 119, no. 16, pp. 59–70, 2018.

[15] R. Van Kranenburg and A. Bassi, "IoT challenges," *Commun. Mob. Comput.*, vol. 1, no. 1, pp. 1–5, 2012.

[16] S. Karmore, R. Bodhe, F. Al-Turjman, R. L. Kumar, and S. Pillai, "Iot based humanoid software for identification and diagnosis of Covid-19 suspects," *IEEE Sens. J.*, vol. 22, pp. 17490–17496, 2020.

[17] J. Gubbi, R. Buyya, S. Marusic, and M. Palaniswami, "Internet of Things (IoT): a vision architectural elements, and future directions," *Fut. Gener. Comput. Syst.*, vol. 29, no. 7, pp. 1645–1660, 2013.

[18] V. Vippalapalli and S. Ananthula, "Internet of Things (IoT) based smart health care system," In *International Conference on Signal Processing, Communication, Power and Embedded System (SCOPES)*, pp. 1229–1233, 2016.

[19] V. S. Rajashekhar and R. Ruban, "Challenges involved in framing additive manufacturing standards," In *Innovations in Additive Manufacturing*, New York, NY: Springer, pp. 321–332, 2022.

[20] A. N. Venkatesh, "Reimagining the future of healthcare industry through Internet of medical things (IoMT), artificial intelligence (AI), machine

learning (ML), big data, mobile apps and advanced sensors," *Artif. Intell. (AI), MAch. Learn. (ML), Big Data, Mob. Apps Adv. Sensors*, 2019.

[21] C. Quispe-Juli, P. Vela-Anton, M. Meza-Rodriguez, and V. Moquillaza-Alcántara, "COVID-19: a pandemic in the age of digital health," *In SciELO Preprints*, 2020. https://doi.org/10.1590/SciELOPreprints.164.

[22] J. P. Tincopa Flores, "Design and evaluation of a diagnostic aid system for diabetic neuropathy based on the reading of plantar pressure points and machine learning," 2019. Available at: https://hdl.handle.net/20.500.12866/7239.

[23] L. Rachakonda, S. P. Mohanty, E. Kougianos, and M. A. Sayeed, "Smart-steering: an IoMT-device to monitor blood alcohol concentration using physiological signals," In *IEEE International Conference on Consumer Electronics (ICCE)*, pp. 1–6, 2020.

[24] I. A. Hashem, I. Yaqoob, N. B. Anuar, S. Mokhtar, A. Gani, and S. U. Khan, "The rise of big date on cloud computing: review and open research issues," *Inf. Syst.*, vol. 47, pp. 98–115, 2015.

[25] D. Davies, "A brief history of cryptography," *Inf. Secur. Tech. Rep.*, vol. 2, no. 2, pp. 14–17, 1997.

[26] M. E. Saleh, A. A. Aly, and F. A. Omara, "Data security using cryptography and steganography techniques," In *IJACSA International Journal of Advances computer Science and Applications*, 2016.

[27] A. Sharma, T. Choudhury, and P. Kumar, "Health monitoring & management using IoT devices in a cloud based framework," In *2018 International Conference on Advances in Computing and Communication Engineering (ICACCE)*, pp. 219–224, 2018.

[28] M. J. Magazine and K. E. Stecke, Research Support Revised April 1994 School of Business Administration.

[29] I. Bonilla-Fabela, A. Tavizon-Salazar, M. Morales-Escobar, T. Guajardo-Muñoz, and C. I. Laines-Alamina, "IoT. The internet of things and the innovation of its applications," *Vinculatégica efan*, vol. 2, no. 1, pp. 2313–2340, 2016.

[30] W. Alhazzni, M. H. Moller, Y. M. Arabi, *et al.*, "Surviving sepsis campaign: guidelines on the management of critically ill adults with Coronavirus Disease 2019 (COVID-19)," *Intensive Care Med.*, vol. 46, no. 5, pp. 854–887, 2020.

[31] R. Akuli, "Internet of Things (IoT) applications with blockchain technique," In *Handbook of IoT and Blockchain, London: CRC Press*, pp. 119–131, 2020.

[32] P. Kumar, G. P. Gupta, and R. Tripathi, "An ensemble learning and fog-cloud architecture-driven cyber-attack detection framework for IoMT networks," *Comput. Commun.*, vol. 166, pp. 110–124, 2021.

[33] S. Al-Sarawi, M. Anbar, K. Alieyan, and M. Alzubaidi, "Internet of Things (IoT) communication protocols," In *8th International Conference on Information Technology (ICIT)*, pp. 685–690, 2017.

[34] X. Yong, D. Chi, and G. Min, "The topology of p2p network," *J. Emerg. Trends Comput. Inf. Sci.*, vol. 3, no. 8, pp. 1213–1218, 2012.
[35] B. Maurer, T. C. Nelms, and L. Swartz, "When perhaps the real problem money itself: the practical materiality of Bitcoin," *Soc. Semiot.*, vol. 23, no. 2, pp. 261–277, 2013.
[36] R. Suomi, R. Cabral, J. F. Hampe, A. Heikkilä, J. Järveläinen, and E. Koskivaara, "Project E-Society: building bricks," In *6th IFIP Conference on E-Commerce, E-Business and E-Government (I3E 2006)*, New York, NY: Springer, 2007.
[37] A. Tapscott and D. Tapscott, *The Blockchain Revolution. Discover How This New Technology Will Transform the Global Economy*, Nueva York: Deusto, 2016.
[38] R. A. Pava-Diaz, J. N. Pérez-Castillo, and L. F. Niño-Vasquez, "Perspective for the use of the P6 model of health care under a scenario supported by IoT and blockchain," *Tecnura*, vol. 25, no. 67, pp. 112–130, 2021.
[39] S. Namasudra, P. Sharma, R. G. Crespo, and S. Vimal, "Blockchain-based medical certificate generation and verification for IoT-based healthcare systems", *IEEE Consumer Electronics Magazine*, 2022. DOI: 10.1109/MCE.2021.3140048.
[40] A. Siddiqui, J. Qaddour, and S. Ullah, "Securing Healthcare IoT (HIoT) monitoring system using blockchain," In *Twelfth International Conference on Ubiquitous and Future Networks (ICUFN)*, pp. 60–66, 2021.
[41] F. Gutiérrez, "A blockchain is a powerful tool against illicit activities: former CIA ex-director," 2021. Online Available: https://www.eleconomista.com.mx/sectorfinanciero/Blockchain-es-una-herramienta-poderosa-contra-actividades-ilicitas-ex-director-de-la-CIA-20210413-0092.html

Chapter 7

Case study: a novel system to manage e-healthcare data by using blockchain technology

Mario Ciampi[1], *Gaia Raffaella Greco*[1], *Fabrizio Marangio*[1,2], *Giovanni Schmid*[1] *and Mario Sicuranza*[1]

Abstract

Changing socio-economic factors, scientific breakthroughs in biomedical research, as well as the emergence of new materials and information technologies are profoundly transforming the healthcare sector. In the last 20 years, a large number of research and development projects and standardization activities have concerned the processing, sharing and presentation of data through reliable, scalable, and economically sustainable systems. In particular, blockchain-based technologies are playing an increasing role in collecting, handling, storing, and updating medical records securely. After the illustration of the current challenges faced by the healthcare organizations in data management, this chapter focuses on a specific use case related to the update of the Patient Summary. This health data document is of fundamental importance for a correct knowledge of the patient's clinical history. The authors developed a blockchain system to support the general practitioner in the Patient Summary update process, ensuring the integrity of the health data and validating the entire medical process through smart contract programming. The proof-of-concept simulation demonstrates the practicability of the technical approach and its ability in improving the current management of the Patient Summary update process.

Keywords: Blockchain; Health information systems; HL7 FHIR; Hyperledger Fabric; Patient Summary; Smart contracts

7.1 Introduction

The advancement of Information and Communication Technologies in the last two decades has made it possible to address many problems in the health sector. In

[1]Institute for High Performance Computing and Networking, National Research Council of Italy, Italy
[2]University of Naples "Parthenope", Italy

particular, healthcare professionals and medical equipment can now produce clinical data and documents in digital format and share them easily, thus supporting physicians in medical diagnosis and reducing possible errors and healthcare costs [1]. Such digital data and documents are managed and stored in several kinds of Health Information Systems (HISs), such as Radiology Information Systems (RISs), Cardiology Information Systems (CISs), Picture Archiving, and Communication Systems (PACs). These systems allow collecting in a structured way data coming from different sources, like medical equipment (Computer Tomography, Magnetic Resonance Imaging, etc.), Electronic Medical Records (EMRs), and Master Patient Indexes (MPIs). The need to allow authorized actors to access a patient's health information quickly enabled the rapid constitution of Electronic Health Record (EHR) systems [2]. Such systems can track patient generated health data available in HISs by collecting a series of related metadata in index registries, pointing to the repositories where this information is stored [3].

Nevertheless, the large amount of data collected over time in HISs does not allow healthcare professionals to efficiently identify the most relevant clinical data to the patients they are treating [4]. For this reason, the General Practitioner (GP) must produce a specific document called Patient Summary (PS), containing the patient's clinical history. This document, unique for each patient, should include the health information that the GP deems more appropriate and should be regularly updated [5]. However, much of this information emerges in clinical visits made by the patient along his/her life (e.g., laboratory or diagnostic reports) and it is reported in other clinical documents. Therefore, the GP could often be not promptly informed [6]. An efficient and secure system able to help GPs in including the new health data in the PS would provide significant support.

Blockchain technology allows a network of users to record data structures in distributed ledgers according to a shared, immutable, and decentralized approach. Although this technology was initially adopted in the financial sector, healthcare providers are rapidly focusing their attention on using it to improve clinical data integrity and quality and securing track healthcare processes [7]. Blockchain technology is typically used together with smart contracts, which are computer programs able to facilitate, verify, and enforce the negotiation or execution of a contract, simulating the logic of contractual clauses [8].

Two circumstances that are radically changing how medical research and practice are conducted and made available to patients in recent times encourage the use of blockchain in healthcare: (i) the growing availability of clinical data sources; (ii) the possibility of managing and sharing this data in digital formats. These factors lead to the production of vast amounts of digital data that require high levels of interoperability and security for their proper processing. Clinical data turns out to be among the types of data most exposed to cyber-attacks because of both their intrinsic value and, due to the spread of wearable medical devices, the significant increase of the attacks.

Interoperability is a crucial factor in correlating data of a heterogeneous nature or produced by different suppliers, obtaining a better understanding of patients' health conditions, and it is also crucial to efficiently retrieve useful information in a vast set of available data and information.

This chapter illustrates a novel system architecture based on blockchain technology and smart contracts that can automatically and securely track the digital health data produced by health professionals and suggest them to the GP of the patient to whom such data refers. In this way, GPs can independently decide whether to insert the information indicated in the PS if they deem it appropriate.

The main contributions described in this chapter regard an application interface and a set of smart contracts that:

- provide support to the GP in editing the PS;
- ensure the reliability and integrity of the overall PS management process;
- improve the quality of the health data with a better trust in health care processes for patients and practitioners.

The chapter is organized as follows. Section 7.2 describes main existing contributions concerning the secure sharing of health information through blockchain technologies. While not directly related to the topic of PS update process, the literature investigation evidences the several use cases where there is an advantage in the use of smart contract features.

Section 7.3 discusses the main blockchain and distributed ledgers applications in the healthcare industry (insurance, supply chain, genomics). Furthermore, the authors in this section overview the fundamental concepts and features of blockchain technologies with particular reference to the Hyperledger Fabric platform used in this work.

The scheme of the proposed system to track digital health data is described in Section 7.4. In particular, the authors illustrate the assets and the rules required at blockchain level and the workflow implemented with the smart contracts through sequence diagrams and pseudo-code.

The authors discuss the results achieved by implementing proof-of-concept simulations illustrated in Section 7.5, where they highlight the main considerations on performance of the system, stressing its practicality and scalability. Finally, Section 7.6 concludes the chapter.

7.2 Literature reviews

Due to its intrinsic interoperability and dependability features and the ability to be event-driven through smart contracts, blockchain is an up-and-coming technology able to address the above data management issues in healthcare. Its introduction in this context is relatively recent, and many projects seem to be still in the embryonic stage. However, there have been many proposals in the last 5 years, primarily around the secure sharing of health information records among stakeholders. In the following, some contributions relevant to this work are described. The reader interested in comprehensive and updated reviews concerning blockchain technology in healthcare can refer to [9–11].

The authors of "*Integrating blockchain for data sharing and collaboration in mobile healthcare applications*" [12] propose a mobile health system for managing data from wearable and non-wearable medical devices. This system aims to improve interoperability and make health data easily accessible to physicians and insurance companies by leveraging blockchain capabilities to ensure privacy and data integrity and improve access control.

The proposal introduced in "*Medrec: Using blockchain for Medical Data Access and Permission Management*" [13] is a decentralized system built on an Ethereum blockchain, used for document management in the EHR. In fact, it permits to ensure accessibility and auditability. The system aims to improve access to medical data, improve interoperability, empower patients to manage their health data by actively participating in the health process, and improve the quality and quantity of health data for research purposes.

With similar goals but for a different context, in "*A blockchain architecture for the Italian EHR system*" [14] and its extension [15], the authors propose a permissioned blockchain architecture designed to manage EHR and ensure the security and integrity of data with a patient-centric approach. The architecture is designed for the Italian public health infrastructure, but can be easily extended in other contexts.

Zhang and White introduce *FhirChain* [16], a blockchain system designed to meet the Office of the National Coordinator for Health Information Technology (ONC) requirements by encapsulating the HL7 Fast Healthcare Interoperability Resources (FHIR) standard [17]. This system predominantly uses blockchain for access control to ensure the security and integrity of clinical data during the various exchanges in the medical field. It has been tested by developing a DApp in the case of clinical data from cancer patients.

The work "*i-Blockchain: A Blockchain-Empowered Individual-Centric Framework for Privacy-Preserved Use of Personal Health Data*" [18] describes blockchain-empowered solutions for the utilization of personal health data through an individual-centric framework that uses an extension of a permissioned blockchain.

In "*Blockchain-Based Medical Certificate Generation and Verification for IoT-based Healthcare Systems*" [19], the authors propose a mobile application that provides an interface between users and healthcare centers to generate and maintain health certificates ensuring privacy and integrity through a blockchain system.

Sharma *et al.* in "*Blockchain-based IoT architecture to secure healthcare systems using identity-based encryption*" [20] propose a blockchain architecture to protect large-scale healthcare data using IBE (Identity Based Encryption) algorithm and a Swarm storage system to reduce the overload.

The proposal "*Blockchain-Powered Parallel Healthcare Systems Based on the ACP Approach*" [21] shows a framework of parallel healthcare systems (PHSs) based on an ACP (Artificial systems plus Computational experiments plus Parallel execution) approach. The blockchain combines with PHS building a consortium to link patients, hospitals, health bureaus, and healthcare communities to improve data sharing, medical records, reviews, and care auditability.

Instead, the authors of "*Healthcare Blockchain System Using Smart Contracts for Secure Automated Remote Patient Monitoring*" [22] propose a blockchain system for real-time analysis of the metadata arising from medical sensors. The system evaluates the data by triggering automated responses through smart contracts to preemptively alert patients and physicians.

The works mentioned so far are mainly on authentication and integrity of the clinical data, without analyzing the health process flow more completely. This circumstance generally applies to the vast body of literature regarding the application of blockchain technology to the healthcare sector.

Therefore, our subsequent studies aim to use blockchain technology to support process authentication. In particular, in the chapter "*Modernizing Healthcare by Using Blockchain*" of "Applications of Blockchain in Healthcare" [23], the authors focus on the current standards used in the healthcare context by devising an architecture for integrating blockchain with some IHE Dynamic Care Profile resources [24]. That system is subsequently refined in *"A blockchain-based smart contract system architecture for dependable health processes"* [25] to fully describe the modules designed by dwelling on a real case of a clinical process.

Unlike some of the proposed works, our proposed architecture does not consider either the insurance context or tokenization (necessary in an environment like Ethereum), thus being disconnected from any economic reasoning not allowed for public health systems in Europe.

Table 7.1 shows the papers analyzed and cited in this work. The Reference column specifies paper's title and reference, Main Goal and Main results columns describe the primary objective of the work and the attained results.

Healthcare is a natively decentralized domain. Indeed, the large number of healthcare facilities distributed throughout the territory daily make clinical, programmatic, and administrative decisions to ensure and improve care provision, in line with the laws and regulations established by national and international authorities. However, some services and databases are managed according to a centralized and client–server approach (e.g., electronic prescription), based on data coming from the territory but often with a single decision point.

As stated in [26], the costs of time and effort required to achieve and sustain cooperative interactions (such as information, negotiation, and execution costs) are limits to centralization. The benefits of adopting a decentralized model in the health sector are numerous: they include the possibility of ensuring the absence of a single point of failure and monopolies; a better development of the health program; transparency; the reduction of errors and duplication of clinical data.

Furthermore, from a technical point of view, as outlined in [27], the client–server paradigm can present some consensus/synchronization problems in distributed systems, such as Byzantine Generals and Sibyl Attack problems.

Blockchain technology can act as a driving force for developing a decentralized model, thanks to its native characteristics. Starting from this, we choose this technology as the basis of the proposed work.

Table 7.1 A list of selected papers related to healthcare data management

Reference	Main goal	Main results
Integrating Blockchain for Data Sharing and Collaboration in Mobile Healthcare Applications [12]	Build a mobile system for collecting personal health data with integrity and privacy	The authors design the system and implement a PoC in *Hyperledger Fabric* for specific use cases, evaluating the system scalability
MedRec: Using Blockchain for Medical Data Access and Permission Management [13]	Realize a comprehensive logging system designed and implemented via *Ethereum* to provide auditability, interoperability and accessibility	*MedRec* was active from January 2018 to September 2019
A Blockchain Architecture for the Italian EHR System[14]	Design an architecture to improve interoperability and ensure data integrity in the Italian national health system	Using *Hyperledger Composer*, the authors implement a PoC application to test the adherence of the smart contracts to the system's functional requirements.
Integrating Blockchain Technologies with the Italian EHR Services [15]	Produce a blockchain layer to solve integrity and traceability issues in the current national EHR framework for the interoperability of the regional systems in Italy	The authors implement a PoC software in *Hyperledger Fabric* to test smart contract functionalities
FHIRChain: Applying Blockchain to Securely and Scalably Share Clinical Data [16]	Create a blockchain system based on the FHIR standard	The system provides patients with more collaborative clinical decision support. The authors test their design through an Ethereum DApp
i-Blockchain: A Blockchain-Empowered Individual-Centric Framework for Privacy-Preserved Use of Personal Health Data [18]	Build framework for managing personal health data based on an extension of permissioned blockchain	The authors propose an architecture and protocols design
Blockchain-Based Medical Certificate Generation and Verification for IoT-based Healthcare Systems [19]	Design a blockchain-based architecture to provide privacy in an IoT-enabled healthcare environment	The authors develop the system in Ethereum and test it using BlockSim
Blockchain-Based IoT Architecture to Secure Healthcare System Using Identity-Based Encryption [20]	Create a privacy-preserving environment for health data that manages medical certificates securely and transparently	The authors develop a PoC and give an evaluation of performance

(Continues)

Table 7.1 (*Continued*)

Reference	Main goal	Main results
Blockchain-Powered Parallel Healthcare Systems Based on the ACP Approach [21]	Develop a prototype for comprehensive healthcare data sharing, medical records review and care auditability	Authors presented a preliminary prototype system called PGDTS and deployed it in China
Healthcare Blockchain System Using Smart Contracts for Secure Automated Remote Patient Monitoring [22]	Use a blockchain-based smart contracts to perform real-time analysis and log transaction metadata for medical sensors in a WBAN	The authors develop a PoC using Ethereum smart contracts
Modernizing Healthcare by Using Blockchain. In Applications of Blockchain in Healthcare [23]	Improve interoperability and validate healthcare processes while at the same time ensuring data integrity by integrating blockchain technology with the HL7 FHIR health informatics standard	The authors develop a PoC through the use of Hyperledger Fabric
A Blockchain-Based Smart Contract System Architecture for Dependable Health Processes [25]	Realize a system to track and verify the implementation of health processes adhering to the HL7 FHIR specifications	The authors develop a PoC for a specific case study using Hyperledger Fabric

7.3 Preliminary studies

7.3.1 The need of a decentralized model for healthcare

Healthcare is a natively decentralized domain. The large number of healthcare facilities distributed throughout the territory daily make clinical, programmatic, and administrative decisions to ensure and improve care provision, in line with the laws and regulations established by national and international authorities. However, some services and databases are managed according to a centralized and client-server approach (e.g., electronic prescription), based on data coming from the territory but often with a single decision point.

As stated in [26], the costs of time and effort required to achieve and sustain cooperative interactions (such as information, negotiation, and execution costs) are limits to centralization. The benefits of adopting a decentralized model in the health sector are numerous: they include the possibility of ensuring the absence of a single point of failure and monopolies; a better development of the health program; transparency; the reduction of errors and duplication of clinical data.

Furthermore, from a technical point of view, as outlined in [27], the client-server paradigm can present some consensus/synchronization problems in distributed systems, such as Byzantine Generals and Sibyl Attack problems.

Blockchain technology can act as a driving force for developing a decentralized model, thanks to its native characteristics. Starting from this, this technology was chosen as the basis of the proposed work.

7.3.2 Blockchain and distributed ledger applications for healthcare

The *Blockchain & Distributed Ledger Observatory of Politecnico of Milano*, established in 2018, publishes annually the report on the state of the art of Blockchain and Distributed Ledgers applications at a global level [28]. In 2020, the study "The Hype is over, get ready for Ecosystems" underlines how the number of initiatives grew by about 59%, while project announcements decreased by 80%.

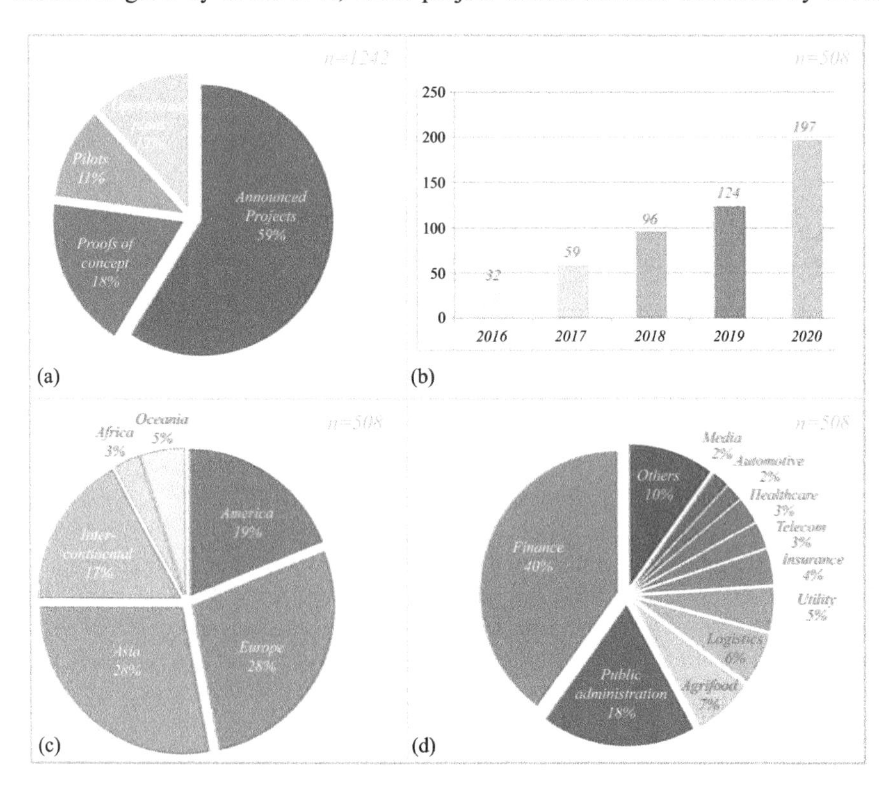

Figure 7.1 Projects on blockchain and distributed ledgers at a global level (2020): (a) projects per level of development; (b) projects per year of establishment; (c) projects per geographic area; (d) projects per sector of application (main). The number of applied projects is 508, as "announced projects" represent 59% of the total (1242). Source: *Politecnico di Milano (2021).*

The shift towards a more mature and concrete market seems confirmed by the constant growth of applicative projects (Figure 7.1(b)) and the increasing number of organizations participating in single projects (systemic approach). As evidenced in Figure 7.1 (a), between 2016 and 2020, announced projects remain the significant majority of the research, development and innovation initiatives investigated in the private and public domains. While intercontinental programs figure as a relevant percentage (17%), Asian and European countries represent more than a half of total applications, showing the biggest number of proof-of-concept, pilots, and operational plans. Figure 7.1(d) describes the weight of the investments in a single industry: financial programs, not surprisingly, represent the biggest economic sector of implementation (almost 40%); public administration projects figure as an 18% of the total, followed by the agrifood sector (7%) and by applications in the logistic industry (6%). As evidenced in the graph, actual proofs-of-concept, pilots, and operational plans in the healthcare industry represent only 3% of the total analyzed. Analysts of the Milan Polytechnic particularly recognize two main trends in the blockchain field. On the *permissionless* side, the increasing availability of non-specific platforms for single applications (open to specific needs and experimentations) as *Ethereum 2.0* and the development of solutions of layer 2. On the other side, general-purpose permissioned platforms are overcoming the experimental phase to reach operability.

These projects are primarily promoted by governments and national or international consortia like the *European Blockchain Services Infrastructure* (*EBSI*), a program participated by the European Union, *Alastria Blockchain Ecosystem*, and *BSN Global* [29]. Although the total number of projects represents a small percentage of the global value, several analysts highlight the relevant economic dimension and the rapid pace of development of the investments of Blockchain and Distributed Ledgers.

Technologies in Healthcare [30,31]. In its report "Blockchain in Healthcare Market: Size, Share, Trends, Growth Opportunity and Forecast 2019–2026" [32], *Acumen Research and Consulting* assesses the global market to reach in 2026 approximately USD 1.7 billion, showing an annual *Compounded Average Growth Rate* (CAGR) of about 48.1%. Following the analyses of the international consulting firm, major forces that will push the growth of the sector are, in the first instance, the increasing adoption of blockchain technologies in the other industries and the need to fasten and secure transactions globally. In the second place, there are the inner dynamics of the health sector. Specifically, the authors describe the growing adoption of digital technologies in health institutions' administrative and medical processes and the growing national average spending per person in healthcare.

The world pandemic has further strengthened these trends. In the last annual report, "Global Life Sciences Outlook" [33], *Deloitte Insight* tries to describe the main trends characterizing the dynamics of health systems, organizations, research, and manufacturing of drugs in the light of the impact of *Covid-19*. Like managers and entrepreneurs interviewed, experts of the consulting firm underline how innovation and novel technologies rapidly entered the healthcare scene.

Applications of blockchain and distributed ledger technologies in healthcare currently span across several fields like supply chain management, clinical trials, insurance, data management (patients, genomics, mobile apps) [34]. Medical supply chain and insurance seem the readiest and most promising use cases. The economic fight against counterfeiting drugs pushes pharmaceutical companies to find systemic and global solutions. At the same time, the general development of blockchain in the insurance industry makes credible a fast adoption also in the health branch [31,35].

Data management – the growing need to collect, handle, organize, communicate, and store patient information – seems one of the trickiest and challenging solutions for medical organizations. Personal, sensitive, and confidential data is often stored across several facilities and providers, resulting in inefficient or inaccessible systems. Distributed Ledgers Technologies offer new tools to store, manage, and update medical records securely, allowing rapid flow across multiple locations, addressing inefficient practices, and preventing data breaches [36]. While applications may seem very promising, many are the wildcards for the healthcare's organizations ecosystem.

Indeed, difficulties stem from poor data handling and security risks. Healthcare data systems are siloed and not always interoperable, thus hindering a wider use and resulting in wasted resources, time, and potential legal exposure given by conflicting or incomplete records.

Johns Hopkins University published in 2016 a breakthrough research on primary causes of death in the United States of America. The investigation showed that the third cause of mortality, after heart disease and cancer, was medical errors (10% of all US Deaths)*. The *National Center for Health Statistics* confirms the importance of medical error as one of the first motivations of death in the USA (200,955 in 2020)†. The fragmentation and untraceability of data across several locations and databases also afflict the European Union countries. This could lead to potential adverse impacts on diagnosis, treatment and follow-up in primary care and community care settings‡. Care approach continuity and updated health records still represent a mission to fulfill.

Bis Research [31] underlines how the information shared among hospitals, physicians, medical organizations, and health providers remains limited also in the most advanced systems worldwide. This dispersion drives rising healthcare costs and is the cause of decreasing quality care. Miscommunication between medical professionals costs an excessive amount of money and is a time-consuming process that creates errors and delays patient care.

As anticipated earlier, the healthcare industry is also plagued by data breaches involving sensitive patient information. *Protenus*, a Baltimore-based healthcare analytics company, publishes annually the *Breach Barometer* report [37]. In 2020, more than 30 million patient records were violated, with an increase of 42% from the previous year. According to the analysts, 62% of healthcare data breaches are

**Johns Hopkins Medicine*, News and Publications (2017). Available: https://www.hopkinsmedicine.org/news/media/releases/study_suggests_medical_errors_now_third_leading_cause_of_death_in_the_us (Accessed on 20 January 2022).

†*National Center for Health Statistics*, Deaths and Mortality (2020). Available: https://www.cdc.gov/nchs/fastats/deaths.htm (Accessed on 18 January 2022).

‡Sam Daley, "*How Using Blockchain in Healthcare Is Reviving the Industry's Capabilities*", 30 July 2021. Available: https://builtin.com/blockchain/blockchain-healthcare-applications-companies (Accessed on 15 January 2022).

hacking incidents, while insider errors were responsible for 20% and 8% were categorized as loss/theft. Unknown causes represent actually 10% of the data breaches, even though the company underlines ransomware as a rising threat to monitor.

Even if different studies exist, analysts stress that the uptake of blockchain and Distributed Ledgers Technologies in the healthcare industry could drive to save up to USD 100 billion per year by 2025 in data breach related costs, IT costs, operations costs, support function and personnel costs, counterfeit related frauds, and insurance frauds [31].

7.3.3 Hyperledger fabric

The purpose of this section is to briefly summarize some of the fundamental concepts and characteristics of blockchain technology, with particular reference to the operating platform used in this work. The reader interested in deepening these topics can refer to the vast literature on the subject, particularly to the recent survey [38] by one of the authors.

Blockchain technology saw its beginnings with the launch of Bitcoin [39] in 2009, and it is now used for many use cases in different application domains. Indeed, many companies are rushing to take advantage of the perceived benefits for business of using a time-oriented, tamper-proof ledger of records as a public or intercompany backbone. The success of blockchain technology mainly lies in demand for interoperable and reliable decentralized architectures, where computers belonging to different and independent administrative domains must cooperate to offer the intended service.

A blockchain is a ledger of transactions replicated among multiple nodes belonging to different trust domains. Each node maintains a copy of the ledger by validating transactions issued by users and grouping them into blocks that include a hash code binding each block to the preceding block. The participating nodes use a consensus protocol to agree on a unique current status of the ledger so that all its replicas are maintained equal and in sync over time.

The properties of the hash function used to build the blockchain make it easy to append new blocks to the current chain but hard to edit, cancel or reorder any block without changing all its subsequent blocks in the chain. Assuming to protect the last current block from tampering, all the blockchain will be. There are two basic ways to get a tamper-resistant blockchain at the current state-of-the-art.

In permissioned blockchain, only correctly identified and authorized nodes can manage the ledger; thus, these nodes can digitally sign valid blocks to protect them from tampering.

In permissionless systems, on the contrary, anyone can participate anonymously in constructing the ledger, and the previous approach is unavailable. The workaround is to make computationally expensive building a new block, getting tamper-resiliency for sufficiently long chains without using digital signatures.

Permissionless systems allow implementing decentralization at a much larger scale than permissioned blockchain but at the cost of much more energy consumption and reduced performance.

Ethereum and Hyperledger Fabric are two relevant examples of permissionless and permissioned systems, respectively. Ethereum and other permissionless systems also allow applications to exchange data only between authorized parties. However, we have preferred to use Fabric to develop our applications in healthcare. Indeed, healthcare processes usually involve a limited number of properly identified participants who have to exchange sensitive information quickly. Furthermore, depending on the scale of the considered application, the system could manage a huge number of processes and participants simultaneously. All this turns out in a system with a low degree of decentralization but excellent transaction latency and throughput performance. Previous experiences in the healthcare domain (e.g., [39,40]) show that Fabric matches these requirements.

Fabric is one of the blockchain projects within the Hyperledger project, established in 2015 by the Linux Foundation to advance cross-industry blockchain technologies. Fabric is an open-source enterprise-grade permissioned blockchain platform with the following relevant features:

- An open governance by a diverse set of maintainers from multiple organizations manages its open design and development.
- It allows for a pluggable set of modules for the membership, ordering, and endorsement services. These services implement the essential functions of access control, consensus protocol, and transaction validation, respectively, thus resulting in a highly modular and configurable architecture.
- Smart contracts can be authored in general-purpose programming languages such as Java, Go, or runtime JavaScript as Node.js.

The last feature stems from the execute-order-validate (EOV) transaction workflow that distinguishes Fabric from other permissioned and permissionless systems. Transactions and the resultant ledger states are two crucial concepts in blockchain technology. A transaction represents the elementary operation tracked in the ledger and the only way to trigger a change in the system. A state is the set of current values stored in the ledger that determines the following possible valid states, along with the transaction rules. In Hyperledger Fabric, a world state database contains the current value of all objects (assets) managed through the system, and a blockchain records the history of all transactions that resulted in the current world state.

In the EOV workflow, a transaction proposal by a client application is first executed by the endorsement service. A subset of peers replicating the ledger (a.k.a. endorsers) realizes this service, where a policy specifies the participants according to the application requirements. If the transaction proposal passes the above validity check, the ordering service assembles it into the next block through a suitable consensus protocol. Finally, all the peers replicating the ledger store the transaction in the current block as valid or invalid, depending on the current state of the ledger. A transaction is valid only if its output, as computed during the endorsement check, matches with the current state of the ledger, and only valid transactions update the ledger state (i.e., the world state database). From the above, it should be clear that the EOV workflow filters out race conditions and other causes

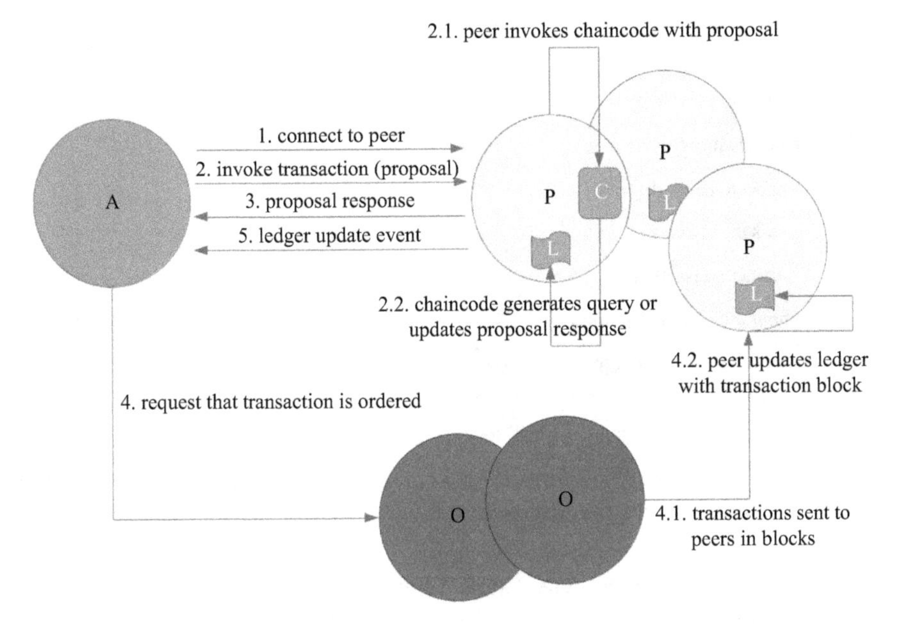

Figure 7.2 The execute-order-validate (EOV) workflow and the components involved. A, P and O denote client application, peers and orderers, respectively. C represents a chaincode and L a blockchain replica.

of non-determinism in asynchronous processing, thus enabling the use of standard programming languages rather than blockchain-specific languages like Script for Bitcoin or Solidity for Ethereum. Figure 7.2 sketches the interactions among the components of a Fabric network for realizing the EOV workflow.

This work mainly concerns developing a decentralized application supporting some healthcare processes through Hyperledger Fabric. Thus, this section concludes by illustrating the basic architecture of a decentralized application in Fabric and the main characteristics of its two building blocks: Fabric SDK and Fabric Chaincode. As shown in Figure 7.2, a decentralized application in Fabric consists of code running on application nodes **A** (a.k.a. Fabric clients) and chaincode running on Fabric peers **P** that act as endorsers.

The term chaincode in Hyperledger Fabric refers to a set of programs and APIs both at the system and application levels. System chaincode is low-level code implementing the various Fabric features described previously. In contrast, application chaincode is used to implement smart contracts enforcing the business processes at the application level. Smart contracts represent the programmed logic of a blockchain system, defining its transactions and how they change the ledger's state. Through any Fabric client with proper permissions, a user application **A** can invoke chaincode to update, query the ledger or invoke another chaincode. Using the chaincode lifecycle, developers and administrators can deploy smart contracts in a Fabric network so that multiple organizations can agree on how those contracts will

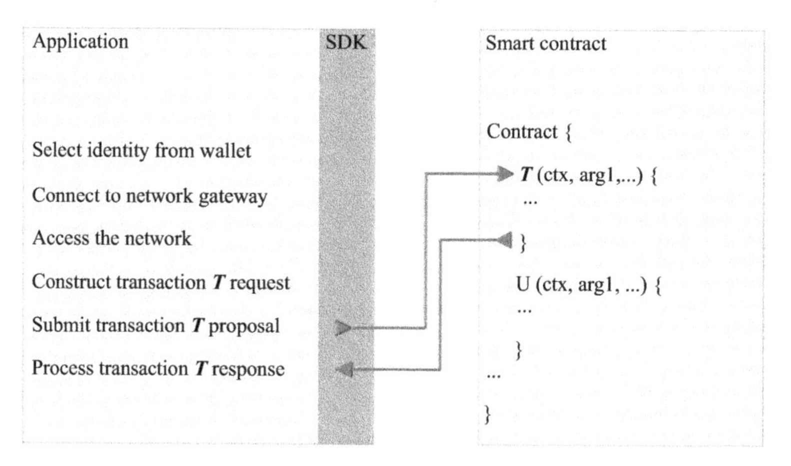

Figure 7.3 The six steps performed by a client application to issue a transaction and retrieve the response

operate before using them. The Fabric Contract API is used to develop smart contracts: application chaincodes must implement the classes and methods in this interface, which allows defining how to interact with the ledger by instantiating suitable transactions. At the time of writing, the Fabric Contract API is provided in Go, Java, and JavaScript.

A client application interacts with a Fabric network using the Fabric SDK. The SDK implements classes and methods to perform the sequence of steps indicated in Figure 7.3.

Diversely than in Bitcoin and other cryptocurrency-based blockchains, a Fabric wallet does not hold any form of cash or token, but a set of X.509 certificates used to access the Fabric network. After selecting an identity from its wallet, the client has to connect to a gateway for identifying one or more peers that provide access to the network. These peers will act as intermediaries between the application and the other nodes of the Fabric network, allowing the client to submit transaction proposals and receive responses through the EOV workflow described above. This section concludes this short overview by underlining that this architecture allows the same client node to connect with multiple blockchain ledgers (a.k.a. Fabric channels), each with its own smart contracts and specific behavior.

7.4 Proposed scheme

The scheme described in this section aims to show how pre-existing healthcare processes can integrate blockchain technology, with the following advantages:

- Easy and effective interoperability of clinical systems and data.
- Support to the clinicians' work.

- Improve the quality of documents and thus the trust of patients and physicians in the system.
- Traceability and validation for the information involved in such processes.

Although any clinical process could integrate the blockchain technology, this work focuses on the Patient Summary and on the possibility of improving the process related to its updating with new clinical data. A proper adoption of the blockchain allows guaranteeing the integrity of the retrieved data and greatly facilitate the task of the General Practitioner, responsible for updating the PS.

The proposed system aims to support general practitioners ensuring the integrity of specific clinical information, focusing in particular on the integrity of the data and the source, thus improving the quality of documents. Moreover, it allows the validation of healthcare processes in which several clinical documents take part in the clinical flow. The current systems prevent the corruption of clinical documents by digitally signing them. Although this methodology effectively preserves the integrity of these documents, it cannot guarantee the integrity of the healthcare process in its entirety. Instead, a distributed system based on blockchain, thanks to its inherent immutability and traceability, is able to guarantee a higher level of security and quality of the processes. The choice of this technology is also extensively justified by an ecosystem intrinsically decentralized and with significant security and interoperability issues. Integrating a single blockchain system could ensure interoperability, strengthen document integrity controls and add functionality related to process' validation. Through smart contracts, it is possible to schedule the flow of transactions that make up healthcare processes, carrying them out securely and reliably. All the operations carried out can be traced back to the users who took part and the assets created, enabling complete and secure audits. The process of updating patient summaries is just one example of how blockchain can fully integrate healthcare processes. This combination would allow for a significant improvement in the quality of the healthcare service provided. In this case, the PS would be updated through a system ensuring the genuineness of information, although under the GP's responsibility.

During the application phase of the rules for collecting information, the system saves the cryptographic hash of this information in the patient's asset. This choice is because the GP may not immediately accept such changes. The proposed system keeps track of the information sent to the GP and, through appropriate checks on the ledger, can verify the consistency with the clinical data present in the source document. All this would not even involve a significant workload for the system since the query operations on the ledger are not subject to endorsement, going to streamline the process. However, it is worth remembering that the responsibility for updating the patient summaries is in charge of the GPs: they must pay attention to the operations to be carried out, despite having a system that supports them in collecting information.

The presented scheme is composed of the integration of the following components:

- HIS represents a system of heterogeneous systems that communicate with each other, including their databases.
- An application layer acts as a communication layer among the HISs, the physicians and the blockchain layer.

- A blockchain client collects data from the application layer and instantiates smart contracts on the blockchain network.
- Smart Contracts are designed around the needs of the process of interest.
- The Ledger is the database storing the smart contracts' artifacts (a.k.a. assets) produced.

As previously described, the PS is a crucial document for proper diagnostic assessment by physicians. This document needs to be up-to-date at all times, which is why it can undergo numerous updates as a result of clinicians' assessments or clinical examinations. The information that populates the PS comes from multiple sources. In general, any clinical exam or specialist visit could result in a PS update. However, its modification is not the responsibility of the physicians who produce the reports. Only the GP will decide whether to add information internally, having full medical and legal responsibility. Therefore, this physician has the arduous task of collecting and verifying patients' health information and entering it into the PS document. Patients hardly voluntarily communicate to the doctor the examinations they have undergone or the results of specialist visits. Consequently, the GP has to request this information from patients, leading to an overload of work and the potential entry of incorrect information.

The schema we have devised aims to support the GP in the management and entry of clinical information through a semi-automated process that guarantees data and process integrity through the use of blockchain. The schema will only propose the information to the doctor, who can decide whether or not to include it.

The creation and updating flows of PS in a generic HIS system and its integration with the blockchain layer are described below.

The flow would start with the transaction of creating a clinical document by a specialist:

1. A physician creates a new clinical document (source document) after a diagnostic examination, visit or other clinical events. The application used by the physician saves the document and its metadata in the HIS.
2. The application used by the GP sends the blockchain client a set of metadata related to the PS needed to initiate the smart contract.
3. A blockchain client launches a transaction to create a blockchain asset for which the source document is issued. The asset stores information such as document type, identifier, and document hash (used for integrity checks) and it is saved on the ledger.
4. After creating the asset, the smart contract checks that the source document is among those whose information could update the PS. For example, the documentation of SARS-CoV-2 vaccination coverage should be included in the PS, specifically in the vaccination section.
5. If the PS needs to be updated, a second transaction fetches the information obtained from the source document. For example, referring to vaccination, the information could be the type of vaccination, the drug used, and the date. All this would be pre-coded in the smart contract, thus guaranteeing the immutability and certainty of the information.

6. The blockchain returns this information to the application level to inform the GP of the latest clinical events concerning his/her patient. The physician, at this point, can choose which information to enter into the PS. For example, suppose that the patient's allergens and sensitivity level remain unchanged after an allergy test. In that case, the physician may decide not to enter this information or only indicate the date of the allergy test. Therefore, the entry is not automated, being the primary responsibility of the GP, but the system essentially support in managing this information.
7. After the GP's digital signature on the updated PS, the blockchain keeps its related updated asset.

Such a system would therefore support GPs and at the same time provides easy access to information. The data in the PS would be aligned and guaranteed by the blockchain processing: indeed, blockchain's inherent immutability characteristics will secure the management of the assets. Although the clinical documents saved in the HIS databases may be subject to attacks or errors, it will always be possible to verify their integrity by comparing their hash values with those saved in their related blockchain assets. In this way, it will always be possible for physicians to verify the integrity of the documents they consult. A scheme of the process described is shown in Figure 7.4. The progressive numbers in the picture reflect the workflow points described above.

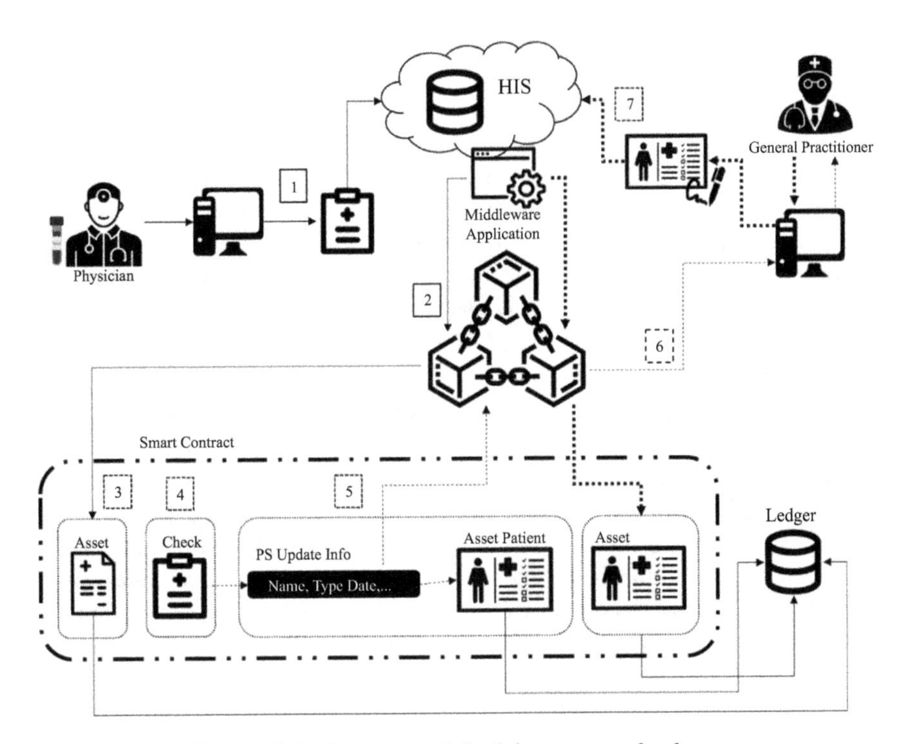

Figure 7.4 System model of the proposed scheme

7.4.1 Assets and rules at the blockchain level

The proper execution of the process previously described is ensured through only two blockchain assets:

- *Clinical document*: this asset results from clinical documents such as prescriptions, diagnostic exams, specialist visits, etc. The asset includes two metadata: the type of document and the hash of the entire clinical document. The first metadata are for ledger searches, while the second metadata allow for integrity checks on the source documents. It is also possible to include other metadata useful for specific searches, such as the patient identifier or the creation date.
- *Patient*: this asset integrates the Patient and Patient Summary concepts at the blockchain level. Compared to the clinical document asset, the structure of this asset is more complex. The metadata in the PS asset are the patient identifier, the hash of the PS, and key-value data structures corresponding to individual sections, indicated by their LOINC code, of the source document. LOINC [41] is a standard coding system for universally identifying medical laboratory observation, used in the context of PS to identify its sections. Each editable section of the Patient Summary is mapped to the asset. Therefore, there are sections related to immunizations, allergies, therapies, etc. For each section, there is a list of entries (e.g., in the immunization section, there is a list of various vaccines performed). There is specific information for each of the above entries, such as source document identifier and date. Each new entry has its identifier and a series of hashes of information related to it. For a vaccine, there could be the name of the vaccine, the date of vaccination, the name of the injected drug, etc. In addition to the hashes, a status field tracks the GP's choices. When the asset is created, the status of the entries is set to *pending*, because only the GP can confirm the entry of any given information. When the GP confirms the insertion in the PS, the status changes to *accepted*, thus recording the operation. The GP may also decide not to enter the information in the PS. In this case, the status is set to *rejected*. In this way, it is possible to track all the information suggested by the system and the choices of the GP.

A transaction called *PsInfoToApp*, encoded in the smart contract, populates the asset with the information extracted from health records. This transaction encodes a set of rules for each source document containing information useful to update the PS document. Clinical documents (usually created in HL7 CDA [42,43] based on XML format) are characterized by a lot of information, much of which is not essential for the update of the PS: e.g., the diversity of codes used to recognize the internal sections of each clinical document. The transaction then extracts the information of interest, useful to the GP for understanding and insertion.

Figure 7.5 shows an example of insertion of data from a pentavalent vaccination document.

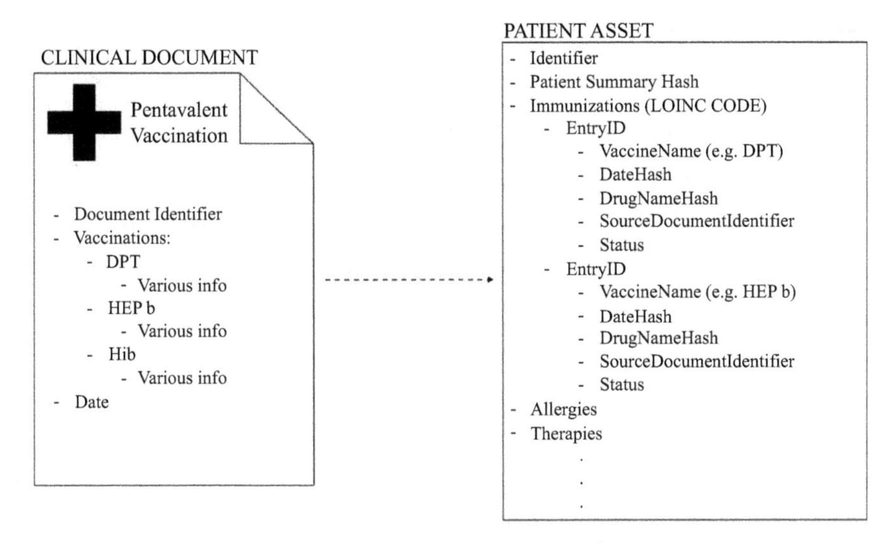

Figure 7.5 Asset creation example

7.4.2 Workflow description

Figure 7.6 shows a sequence diagram related to a complete execution to understand the process better. The numbers present in this figure are relative to the sequential numbers of Figure 7.4. This type of diagram shows how operations are carried out and the interactions among the components, temporally ordering the individual steps. It is assumed that the system recognizes the source clinical document as valid for extracting information and that the GP accepts the proposed changes.

For the sake of clarity, all the operations described below in order of execution are listed below.

- **Create&Update_Client:** *CreateDocument*
  A physician submits a document creation task through the application interface. After creating the document on an appropriate HIS database, the client sends the information to the *MiddlewareApplication*. In the meantime, the HIS sends confirmation of the correct creation of the document on its database.
- **Middleware Application:** *CreateDocument*
  The *Application* notifies the *Fabric Client* about the creation of a new document, sending it the metadata needed to execute the appropriate transaction.
- **Fabric Client:** *CreateClinicalDocument*
  Upon receiving the *Application's* request, the *Fabric Client* instantiates the smart contract and submits to the endorsers the transaction proposal related to the creation of a new *ClinicalDocument* asset on the blockchain. After receiving a positive response from the endorsers, the Fabric Client provides the Smart Contract with some metadata and requests the submission of the transaction.

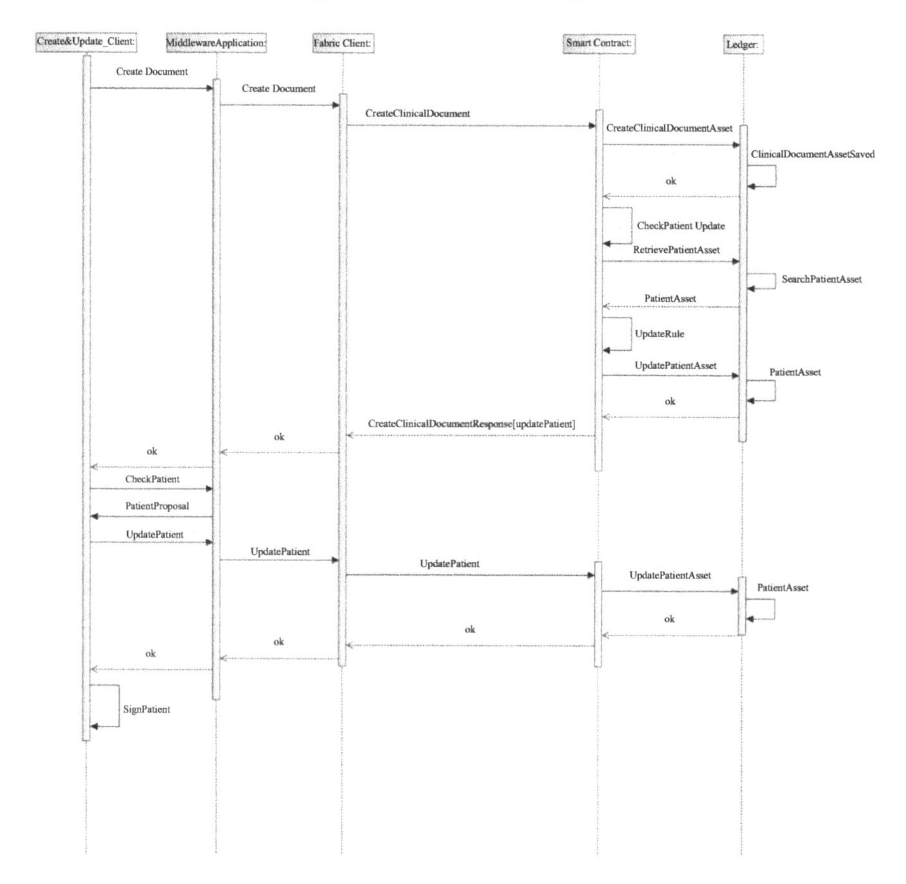

Figure 7.6 Sequence diagram

- **Smart Contract:** *CreateClinicalDocumentAsset*
  The smart contract creates a new asset for the clinical document and saves it on the *Ledger*. In case of errors, the relevant codes are returned, otherwise the transaction continues.
- **Smart Contract:** *CheckPatientUpdate*
  The smart contract verifies that the created document type (identified by its LOINC code) is present in the list of documents that can update the Patient Summary. If not, the smart contract stops processing; otherwise, it goes to the next step.
- **Smart Contract:** *RetrievePatientAsset*
  The smart contract uses the Patient ID to read the Patient asset from the ledger.
- **Smart Contract:** *UpdateRule*
  After receiving the Patient asset, the smart contract applies the Update rules for the Patient Summary required for the source document. It also identifies the sections of the Patient Summary to be updated.

- **Smart Contract:** *UpdatePatientAsset*
  The smart contract updates the Patient asset as established in the previous step, setting to "*pending*" the status for each suggested record, and saves it on the Ledger. The status will subsequently vary depending on the choices made by the GP.
- **Contract:** *CreateClinicalDocumentResponse*
  The smart contract returns the information records and the sections of the Patient Summary where these records should be added to the Fabric Client. The Fabric Client then sends an Update proposal to the application, which initiates the *UpdatePatient* action by interacting with the client.
- **Fabric Client:** *UpdatePatient*
  The application sends the GP decision to the Fabric Client, indicating which information will actually update the Patient Summary, thus updating the *status* of the PS sections in the Patient asset. The hash of the new updated Patient Summary document is also sent.
- **Smart Contract:** *UpdatePatientAsset*

Based on the GP's decision, the smart contract requests a new update of the Patient asset. This update modifies the status associated with the previously created sections and the PS hash value, thus realigning the PS to the version just updated by the GP (the hash must not consider any information in "*rejected*" status). After successfully updating the asset, the smart contract returns to the application, which in turn shows the asset to the client.

The following pseudocode summarizes the various steps.

`APPLICATION LEVEL Step 1`

```
create clinicalDocument;
send clinicalDocument to HIS application;
{
In the HIS application
take StreamInput;
save as Clinical_Document in database;
    send save confirmation to requestor;
}
send clinicalDocument to Middleware application;
{
In the Middleware application
take StreamInput;
Obtain metadata;
send document and metadata to Fabric Client;
}
```

BLOCKCHAIN LEVEL Step 1

```
create Clinical_Document_Asset;
save Clinical Document Asset on Ledger;
if (Clinical Document Type == Valid){
        retrieve Patient_Asset;
        apply predefined rules;
        extract information;
        update patient_Asset;
        return update_proposal;
} else {
        return "process complete";
}
```

APPLICATION LEVEL Step 2

```
if (return == "update proposal")
    {
    show proposal to screen //(for the gp)
    wait for gp_decision ;
    // wait general practitioner decision
    if(gp_decision == yes)
    {
    send update_request ;
    }
    else {return "exit";}
}
```

BLOCKCHAIN LEVEL Step 2

```
if (exists(update request)
    {
    update Patient_Asset;
    send confirmation to application level;
    }
```

APPLICATION LEVEL Step 3

```
sign new Patient_Summary
send Patient_Summary in HIS database
```

7.5 Results and discussion

The proposed system aims to support general practitioners in updating the patient summary, a pivotal document for an immediate knowledge of the patient's clinical history. Indeed, this document collects all health information concerning a given individual, supporting physicians in a variety of medical situations. Unfortunately, without controlled automation of the process, general practitioners may lack the necessary information or receive it late. For instance, it is possible to consider a patient who goes to a medical center for an allergy testing and an allergy to a drug is diagnosed. Ideally, this information should be transmitted automatically from the testing center to the primary care physician: if this transmission fails, the patient should directly inform his/her physician. In reality, these steps often do not occur or occur late. If the patient was to be hospitalized in an unconscious state, emergency room physicians might inoculate the wrong medications due to lack of information.

From this simple scenario, it is clear the need to update the patient summary quickly and efficiently. The system proposed in this chapter allows not only to support the primary care physician by automating the process, but exploits the potential of the blockchain so as to ensure the integrity of the data included in the patient summary.

7.5.1 Experimental environment

An experimentation of the proposed system has been carried out on a Hyperledger Fabric v. 2.4 test network to verify the proper functioning of the implemented smart contracts. The network was simulated on an Ubuntu 20.04 VM, designing the smart contracts through the IBM Blockchain Platform Developer Tools [44] installed as a VS Code extension. The simulation of the application layer has been done through Java scripts, whereas the communication with the blockchain has been implemented through the Hyperledger Fabric Java SDK [45]. A clinical document referring to the patient identified through the fiscal code (unique identifier used in Italy in the health domain) "RSSMRA83M11H501P" to carry out the tests was generated. The clinical document (a.k.a. source document) was a vaccination document concerning the single inoculation of Pfizer's COVID-19 vaccine. For this reason, the section of the PS to be updated concerns the "Immunization" section, identified in the case under examination by the LOINC code 11369-6.

7.5.2 Proof of concept simulations

The evolution of health documents used during the simulations is illustrated below. The first two boxes show two extracts from the PS represented according to the HL7 CDA format concerning the "Immunization" section, before being modified by the system.

```
<code code="11369-6" codeSystem="2.16.840.1.113883.6.1"
codeSystemName="LOINC"
      displayName="History of immunization"/>
<title>Immunizations</title>
<text>
   <table>
      <thead>
         <tr>
         <th>Vaccine</th>
         <th>Date</th>
         <th>State</th></tr></thead>
   <tbody>
      <tr>
         <td>
            <content ID="IMMUNIZATIONS_1">Inflexal V
                    </content></td>
         <td>december 2011</td>
         <td>completed</td></tr>
      <tr>
         <td>
            <content ID="IMMUNIZATIONS_2">Flu Vaccine, Split
             Virus or Surface Antigen</content></td>
         <td>november 2013</td>
         <td>completed</td></tr></tbody></table></text>
```

The first snippet represents the section summarizing the various entries performed over time: the name, date, and status of each vaccine are present. This data represents basic information on inoculations.

```
<entry>
   <substanceAdministration classCode="SBADM" moodCode="EVN"
     negationInd="false">
      <templateId root="1.3.6.1.4.1.12559.11.10.1.3.1.3.19"/>
      <templateId root="1.3.6.1.4.1.19376.1.5.3.1.4.12"/>
      <templateId root="2.16.840.1.113883.10.20.1.24"/>
```

```
        <id root="48E1A5DC-E9C4-4458-A454-F3D5D45E5555"/>
        <code code="IMMUNIZ" codeSystem="1.3.5.1.4.1.
            19376.1.5.3.2" codeSystemName="IHEActCode"/>
        <text>
            <reference value="#IMMUNIZATIONS_1"/>
        </text>
        <statusCode code="completed"/>
        <effectiveTime value="201112"/>
        <consumable typeCode="CSM">
          <manufacturedProduct classCode="MANU">
            <templateId root="1.3.6.1.4.1.12559.11.10.1.3.1.3.18"/>
            <templateId root="1.3.6.1.4.1.19376.1.5.3.1.4.7.2"/>
            <templateId root="2.16.840.1.113883.10.20.1.53"/>
            <manufacturedMaterial classCode="MMAT" determiner
              Code="KIND">
                <code code="721863005" codeSystem
                  ="2.16.840.1.113883.6.96" codeSystemName="SNOMED-
                  CT" displayName="Antigen of Influenza virus surface
                  protein (substance)">
                  <originalText>
                      <reference value="#IMMUNIZATIONS_1"/>
                  </originalText>
                  <translation codeSystem="2.16.840.1.113883.
      2.9.6.1.5" codeSystemName="AIC" displayName="Inflexal V"/>
                      </code>
                      <name>Inflexal V</name>
                </manufacturedMaterial>
            </manufacturedProduct>
        </consumable>
    </substanceAdministration>
</entry>
```


The second snippet represents the section of the PS specific to a given vaccine (in this case Inflexal V). Leaving aside the considerable amount of codes present and not essential for the discussion, this section contains more specific information regarding inoculation. It is possible noting the presence of the AIC code (an Italian coding system for identifying medicinal products), the more detailed description of the type of vaccine administered, and the date field.

A physician initiates the process flow by creating the clinical document in the HIS (in the example shown, the clinical document is a vaccination document). The following box shows an extract of the vaccination document, highlighting the critical information that is managed by the process to eventually update the PS, such as the date, the name of the vaccine, and the AIC code.

```
<substanceAdministration classCode="SBADM" moodCode="EVN">
  <templateId root="2.16.840.1.113883.2.9.10.1.11.4.1" />
  <statusCode code="completed" />
  <effectiveTime value="20210608183209+0100" />
  <routeCode code="IVPUSH" codeSystem="2.16.840.
   1.113883.1.11.14628" codeSystemName="v3-RouteOfAdministra-
   tion" displayName="Injection, intravenous, push" />
  <approachSiteCode codeSystem="2.16.840.1.113883.6.262" code-
   SystemName="v3-ActSite" code="LUA" displayName="left upper
   arm" />
  <doseQuantity value="123" unit="mg" />
  <consumable typeCode="CSM">
   <!- Injected vaccine ->
   <manufacturedProduct>
    <manufacturedMaterial>
      <code>049269018</code>
      <!- AIC Code ->
      <lotNumberText>1515</lotNumberText>
      <displayName>COMIRNATY-Pfizer- BioNtech</displayName>
      </manufacturedMaterial>
      <manufacturerOrganization>
        <id />
        <name />
      </manufacturerOrganization>
    </manufacturedProduct>
  </consumable>
</substanceAdministration>
```

Figure 7.7 shows the Patient asset without displaying the metadata content related to the PS sections to simplify visualization.

Metadata represent a fillable section of the PS; it is easy to note that, unless there is a minor manipulation due to programming requirements, the name given to the various sections links directly to the relative LOINC codes. Thus, the metadata related to the hash of the PS and to the immunization section characterized by its LOINC code can easily be identified.

Following the creation from the physician, the smart contract creates a "clinical document" asset on the blockchain (Figure 7.8).

This asset contains a hash of the document and its type, identified by its LOINC code. Figure 7.9 shows the resulting asset.

After creating the "clinical document" asset, the smart contract verifies that the type of clinical document created requires a PS update. If so, the smart contract proceeds by identifying the specific rules for the given clinical document, fetching the necessary information and the section of the PS where to save this information.

A proposed patient summary update is then be constructed and sent to the GP. Figure 7.9 shows the values of the patient's asset fields, highlighting the pending status: information that has not yet been evaluated by the GP (accepted or rejected)

```
{} Patient
   patientSummaryHash : "f9d2084ec2447bcd51857cf4560aea72278068c790f0ffeb903a264472cd9507"
{} c11369_6
{} c48765_2
{} c10160_0
{} c47420_5
{} c46264_8
{} c47519_4
{} c8716_3
{} c29762_2
```

Figure 7.7 Patient asset

```
{} ClincalDocument
   hash : "c264fb16caa6e9fce294154cca19697e04d79d78b75bb01aa057d9e126d1688c"
   type : "87273-9"
```

Figure 7.8 ClinicalDocument asset

```
{} Patient
   patientSummaryHash : "f9d2084ec2447bcd51857cf4560aea72278068c790f0ffeb903a264472cd9507"
   {} c11369_6
      {} f851c3faba66ae11f3ecb7e52eacd6359c21c93d15ac5abc61fec6b8483edd99
         Status : "pending"
         Source Document ID : "TEST_CLINICAL_DOCUMENT"
         Status Code Hash : "4ddb3e96801a1ee2b77dc5247c0db478d5f97a93b90e7cdb09f5f51d43764b08"
         Effective Time Value Hash : "821b61801683f4d3782c745709f7967ae63aa46e80034dbcb71670e1954d2b64"
         Aic Code Hash : "241c79a63db3e4df078a55dfe3d19a92aefc4875027a41d265d026b9129ca7bf"
         Patient Summary Section Hash : "2143e4ee10e0bf9b324596ea5e6573d4a54bcebb67e4e34a27794ea12b3e9921"
         Display Name Hash : "9bd10bff06313a80f50c2b102828ac1f77ff8f8593feb7466b8c504b3b755bcf"
```

Figure 7.9 Patient asset after information entry

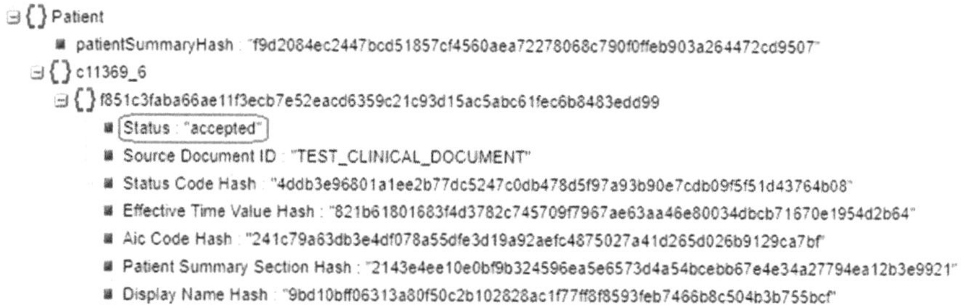

```
{} Patient
    patientSummaryHash : "f9d2084ec2447bcd51857cf4560aea72278068c790f0ffeb903a264472cd9507"
  {} c11369_6
    {} f851c3faba66ae11f3ecb7e52eacd6359c21c93d15ac5abc61fec6b8483edd99
        Status : "accepted"
        Source Document ID : "TEST_CLINICAL_DOCUMENT"
        Status Code Hash : "4ddb3e96801a1ee2b77dc5247c0db478d5f97a93b90e7cdb09f5f51d43764b08"
        Effective Time Value Hash : "821b61801683f4d3782c745709f7967ae63aa46e80034dbcb71670e1954d2b64"
        Aic Code Hash : "241c79a63db3e4df078a55dfe3d19a92aefc4875027a41d265d026b9129ca7bf"
        Patient Summary Section Hash : "2143e4ee10e0bf9b324596ea5e6573d4a54bcebb67e4e34a27794ea12b3e9921"
        Display Name Hash : "9bd10bff06313a80f50c2b102828ac1f77ff8f8593feb7466b8c504b3b755bcf"
```


Figure 7.10 Patient asset after positive GP choice

is managed in this way to facilitate further checks or audits. Note that Figure 7.9 displays only the entry related to the current vaccine: for clarity reasons, the other existing PS entries and sections are omitted.

After that, when the GP checks for any update suggestions for the patient, the system suggests that the "Immunization" section could be updated. The GP can then choose whether or not to update the PS with this information; in affirmative case, the PSupdate service on the HIS is started and the patient's asset in the BC is updated. As shown in the following two code boxes, the PS will have a new section regarding the new vaccination, while the status of the asset will be changed to "accepted" (Figure 7.10).

```
<code code="11369-6" codeSystem="2.16.840.1.113883.6.1" codeSys-
   temName="LOINC"
       displayName="History of immunization"/>
<title>Immunizations</title>
<text>
    <table>
      <thead>
        <tr>
          <th>Vaccine</th>
          <th>Date</th>
          <th>State</th></tr></thead>
      <tbody>
        <tr>
          <td>
             <content ID="IMMUNIZATIONS_1">Inflexal V</content></td>
          <td>december 2011</td>
```

```
        <td>completed</td></tr>
    <tr>
        <td>
        <content ID="IMMUNIZATIONS_2">Flu Vaccine, Split Virus or
    Surface Antigen</content></td>
        <td>november 2013</td>
      <td>completed</td></tr></tbody></table></text>
        <content ID="IMMUNIZATIONS_3">COMINRATY-Pfizer-
                BioNtech</content></td>
      <td>June 2021</td>
      <td>completed</td></tr></tbody></table></text>
<entry>
   <substanceAdministration classCode="SBADM" moodCode="EVN"
        negationInd="false">
      <templateId root="1.3.6.1.4.1.12559.11.10.1.3.1.3.19"/>
      <templateId root="1.3.6.1.4.1.19376.1.5.3.1.4.12"/>
      <templateId root="2.16.840.1.113883.10.20.1.24"/>
      <id root="48E1A5DC-E9C4-4458-A454-F3D5D45E5555"/>
      <code code="IMMUNIZ" codeSystem="1.3.5.1.4.1.
        19376.1.5.3.2" codeSystemName="IHEActCode"/>
      <text>
        <reference value="#IMMUNIZATIONS_3"/>
      </text>
      <statusCode code="completed"/>
      <effectiveTime value="20210608183209+0100"/>
      <consumable typeCode="CSM">
        <manufacturedProduct classCode="MANU">
          <templateId root="1.3.6.1.4.1.12559.11.10.1.3.1.3.18"/>
          <templateId root="1.3.6.1.4.1.19376.1.5.3.1.4.7.2"/>
          <templateId root="2.16.840.1.113883.10.20.1.53"/>
          <manufacturedMaterial classCode="MMAT"
            determinerCode="KIND">
             <code code="049269018" codeSystem="2.16.840.1.
               113883.2.9.6.1.5" codeSystemName="AIC" display-
               Name="COMINRATY-Pfizer- BioNtech">
                <originalText>
```

```
                    <reference value="#IMMUNIZATIONS_3"/>
                </originalText>
                </code>
                <name>COMINRATY-Pfizer- BioNtech</name>
            </manufacturedMaterial>
        </manufacturedProduct>
    </consumable>
</substanceAdministration>
</entry>
```

7.5.3 Performance considerations

The system described fulfills the purpose of the work, providing correct results consistent with expectations. Of course, the integration with the blockchain requires to deal with the performance of a complex network that, in healthcare, must be able to handle large data streams and numerous peers. While early blockchain systems tended to store raw data, newer architectures select only the metadata of interest to avoid scalability problems. These considerations are behind the use of Hyperledger Fabric as a framework. First, the type of blockchain it offers is related to the concept of consortium and disconnected from tokenization, choices that well match the healthcare context. Second, Fabric's performance appears to be adequate for the workloads of the context of interest.

Table 7.2 shows statistics on drug prescriptions in 2019 in Italy [46]. Although it is only a part of the healthcare processes that daily occur, this table gives an idea of the data flows in the healthcare context. Considering each process as a transaction and assuming 20 working hours, it is possible to theorize about 10,000 transactions per hour, equivalent to about 2.78 transactions per second due to hospitals' continuous hours. Although this number of transactions is an estimate, there would be no significant change in theoretical performance even by significantly increasing this number. Recent studies on the Hyperledger Fabric performance attest the ability of this framework to support thousands of transactions per second. In [47], the authors carried out a performance analysis of Hyperledger Fabric when varying the type of transactions (read or write) and considering the application of Bloom Filtering or compression. The results show a number of transactions per second between 2,000 and 6,000. In general, read transactions have significantly less weight than write transactions, a characteristic easily attributable to the lack of writing information to the ledger. So, the number of transactions, while varying depending on the type and filters applied, remains very high and in line with the context of interest.

Table 7.2 Statistics on drug prescriptions in Italy

Subject	Per years	Details
Drugs	2.2 million treatments	72% National Health Service treatments served
People served	4.6 million children and adolescents 1.9 million new patients	1,600 servings per 1,000 inhabitants in a day
Prescription healthcare processes	About 1,600,000	5,000 processes a day about 500 processes in a hour

Given the variability related to the type of transaction considered, generic performance studies cannot fully validate the proposed architecture. Also, considerable changes may occur at the performance level depending on the designed network's morphology. The number of peers, the type of Certification Authorities, the consensus mechanism, and the number of orderers and endorsers could affect the system's performance to a greater or lesser extent. In this respect, it is essential to note the importance of the choice of endorsement policies. Hyperledger Fabric tends to use a basic majority endorsement policy, in which a majority of the organizations participating in the network must express consent for a given transaction. In the healthcare context, a multiplicity of actors could be mapped as organizations within the network, like hospitals, clinics, pharmacies, testing centers, and so on. It is therefore critically important that the network be structured to minimize the number of endorsers or that endorsement policies be used to minimize the computational burden, measured by proper tools like Hyperledger Caliper [48].

7.6 Conclusion

This chapter investigates the application of Distributed Ledger Technologies and, more in particular, of Blockchain technology to improve the quality and integrity of healthcare processes. Such technologies are rapidly spreading across several application domains due to their ability to foster decentralization by immutably tracking all transactions performed in a consortium composed of peer participants.

The nature of the healthcare domain, characterized by numerous health professionals and independent hospitals, medical laboratories, ambulatories, GP offices, as well as heterogeneous clinical data and documents produced by the multiple authorized health actors, makes it a natural sector where these technologies can provide a significant support.

The chapter examines these aspects and presents a novel scheme able to integrate the Blockchain technology with the healthcare processes to collect the most significant health data generated by Health Information Systems, in order to support General Practitioners (GPs) to create and update the Patient Summary. The Patient Summary is a document created and updated by a GP, which summarizes the clinical history of a patient. More in detail, the scheme proposed in this chapter, based on the adoption of both blockchain technology and smart contracts, aims to facilitate GPs to be aware of all the new clinical data produced for their patients, in order to make them able to decide whether include this information in the document. Indeed, this document is particularly important in the health domain, as it contains the primary data concerning a patient's clinical history. For this reason, Patient Summary needs particular attention relating to the integrity of its content and production process.

The system proposed in this chapter is able to satisfy both the needs through the design and use of ad-hoc smart contracts. A proof-of-concept system was implemented using Hyperledger Fabric. This framework provides consolidated performances that permit to respect the strict requirements of the main health processes. It is worth noting that the proposed system can more generally be used for various health documents and processes. This aspect underlines that blockchain is an enabling technology of considerable interest in the healthcare domain.

As future work, the authors intend to evaluate in a more real scenario the proposed system architecture. In particular, once the smart contracts are tested, their proper functioning will be verified on an opportunely designed network in order to evaluate the final performances. To this end, specific measurements are already being acquired through a testbed using Hyperledger Caliper.

References

[1] A. Ariani, A. P. Koesoema, and S. Soegijoko, "Innovative healthcare applications of ICT for developing countries," in H. Qudrat-Ullah and P. Tsasis, Eds., *Innovative Healthcare Systems for the 21st Century, Understanding Complex Systems*, Springer, Cham, 2017, pp. 15–70.

[2] S. V. Jardim, "The electronic health record and its contribution to healthcare information systems interoperability," *Procedia Technology*, vol. 9, pp. 940–948, 2013.

[3] M. Ciampi, M. Sicuranza, A. Esposito, R. Guarasci, and G. De Pietro, "A technological framework for EHR interoperability: experiences from Italy," in C. Röcker, J. O'Donoghue, M. Ziefle, M. Helfert, and W. Molloy, Eds., *Information and Communication Technologies for Ageing Well and e-Health, Communications in Computer and Information Science*, vol. 736, Springer, Cham, 2016, pp. 80–99.

[4] C. Zhang, R. Ma, S. Sun, Y. Li, Y. Wang, and Z. Yan, "Optimizing the electronic health records through big data analytics: a knowledge-based view," *IEEE Access*, vol. 7, pp. 136223–136231, 2019.

[5] eHealth Network, Guideline on the Electronic Exchange of Health Data under Cross-Border Directive 2011/24/EU: Patient Summary. Available: https://ec.europa.eu/health/sites/default/files/ehealth/docs/ehn_guidelines_patientsummary_en.pdf, 2021 [Accessed on 31 July 2022].

[6] European Commission. Electronic Cross-Border Health Services. Available: https://ec.europa.eu/health/ehealth/electronic_crossborder_healthservices, 2022 [Accessed on 31 July 2022].

[7] A. Shahnaz, U. Qamar and A. Khalid, "Using blockchain for electronic health records," *IEEE Access*, vol. 7, pp. 147782–147795, 2019.

[8] S. Wang, L. Ouyang, Y. Yuan, X. Ni, X. Han, and F.-Y. Wang, "Blockchain-enabled smart contracts: architecture, applications, and future trends," *IEEE Transactions on Systems, Man, and Cybernetics: Systems*, vol. 49, no. 11, pp. 2266–2277, 2019.

[9] A. Hasselgren, K. Kralevska, D. Gligoroski, S. A. Pedersen, and A. Faxvaag, "Blockchain in healthcare and health sciences – a scoping review," *International Journal of Medical Informatics*, vol. 134, p. 104040, 2020.

[10] A. Tandon, A. Dhir, A. N. Islam, and M. Mäntymäki, "Blockchain in healthcare: a systematic literature review, synthesizing framework and future research agenda," *Computers in Industry*, vol. 122, p. 103290, 2020.

[11] S. Namasudra and G. D. Chandra, *Applications of Blockchain in Healthcare,* Springer, New York, NY, 2021.

[12] X. Liang, J. Zhao, S. Shetty, J. Liu, and D. Li, "Integrating blockchain for data sharing and collaboration in mobile healthcare applications," in *Proceedings of the 2017 IEEE 28th Annual International Symposium on Personal, Indoor, and Mobile Radio Communications*, IEEE, Montreal, QC, Canada, 2017.

[13] A. Azaria, A. Ekblaw, T. Vieira, and A. Lippman, "MedRec: using blockchain for medical data access and permission management," in *Proceedings of the 2016 2nd International Conference on Open and Big Data*, IEEE, Vienna, Austria, 2016.

[14] M. Ciampi, A. Esposito, F. Marangio, G. Schmid, and M. Sicuranza, "A blockchain architecture for the Italian EHR system," in *Proceedings of the Fourth International Conference on Informatics and Assistive Technologies for Health-Care, Medical Support and Wellbeing, ThinkMind Digital Library*, Valencia, Spain, 2019.

[15] M. Ciampi, A. Esposito, F. Marangio, G. Schmid, and M. Sicuranza, "Integrating blockchain technologies with the Italian EHR services," *International Journal on Advances in Life Sciences*, vol. 12, pp. 57–69, 2020.

[16] P. Zhang, J. White, D. C. Schmidt, G. Lenz, and S. T. Rosenbloom, "FHIRChain: applying blockchain to securely and scalably share clinical data," *Computational and Structural Biotechnology Journal*, vol. 16, pp. 267–278, 2018.

[17] HL7 Fast Healthcare Interoperability Resources. Available: https://www.hl7.org/fhir/, 2022 [Accessed on 31 July 2022].

[18] K. Ito, K. Tago, and Q. Jin, "i-Blockchain: a blockchain-empowered individual-centric framework for privacy-preserved use of personal health data," in *Proceedings of the 2018 9th International Conference on Information Technology in Medicine and Education*, IEEE, Hangzhou, China, 2018, pp. 829–833.

[19] S. Namasudra, P. Sharma, R. G. Crespo, and V. Shanmuganathan, "Blockchain-based medical certificate generation and verification for

IoT-based healthcare systems," *IEEE Consumer Electronics Magazine*, pp. 2162–2256, 2022.

[20] P. Sharma, N. R. Moparthi, S. Namasudra, V. Shanmuganathan, and C.-H. Hsu, "Blockchain-based IoT architecture to secure healthcare systems using identity-based encryption," *Expert Systems*, vol. 39, no. 10, e12915, 2021.

[21] S. Wang, J. Wang, X. Wang, *et al.*, "Blockchain-powered parallel healthcare systems based on the ACP approach," *IEEE Transactions on Computational Social Systems*, vol. 5, no. 4, pp. 942–950, 2018.

[22] K. N. Griggs, O. Ossipova, C. P. Kohlios, A. N. Baccarini, E. A. Howson, and T. Hayajneh, "Healthcare blockchain system using smart contracts for secure automated remote patient monitoring," *Journal of Medical Systems*, vol. 42, no. 7, pp. 1–7, 2018.

[23] M. Ciampi, A. Esposito, F. Marangio, G. Schmid, and M. Sicuranza, "Modernizing healthcare by using blockchain," in S. Namasudra and G. D. Chandra, Eds., *Applications of Blockchain in Healthcare*, Springer, Cham, 2021, pp. 29–67.

[24] IHE Profiles. Available: https://www.ihe.net/resources/profiles/, 2022 [Accessed on 31 July 2022].

[25] M. Ciampi, F. Marangio, G. Schmid, and M. Sicuranza, "A blockchain-based smart contract system architecture for dependable health processes," in *Proceedings of the Italian Conference on CyberSecurity*, CEUR-WS, vol. 2940, 2021.

[26] S. Abimbola, L. Baatiema, and M. Bigdeli, "The impacts of decentralization on health system equity, efficiency and resilience: a realist synthesis of the evidence," *Health Policy and Planning*, vol. 34, no. 8, p. 605–617, 2019.

[27] T.-T. Kuo and L. Ohno-Machado, "ModelChain: decentralized privacy-preserving healthcare predictive modeling framework on private blockchain networks," *arXiv:1802.01746v1,* 2018.

[28] P. di Milano, Osservatorio Blockchain e Distributed Ledgers, Report, "Blockchain – The Hype is over, Get Ready for Ecosystems, Infographics. Available: https://www.osservatori.net/it/ricerche/infografiche/blockchain-hype-is-over-get-ready-ecosystems-infografica, 2021 [Accessed on 31 July 2022].

[29] P. Soldavini, La blockchain cresce per ecosistemi concentrandosi sui progetti operativi, Sole24Ore. Available: https://www.ilsole24ore.com/art/la-blockchain-cresce-ecosistemi-concentrandosi-progetti-operativi-ADgLETEB, 21 February 2021 [Accessed on 31 July 2022].

[30] Mordor Intelligence, Blockchain Market in Healthcare – Growth, Trends, Covid-19 Impact, and Forecasts (2021–2026). Available: https://www.mordorintelligence.com/industry-reports/blockchain-market-in-healthcare, 2020 [Accessed on 31 July 2022].

[31] BIS Research, Global Blockchain in Healthcare Market. Available: https://bisresearch.com/industry-report/global-blockchain-in-healthcare-market-2025.html, 2018 [Accessed on 31 July 2022].

[32] Acumen Research and Consulting, Blockchain in Healthcare Market: size, Share, Trends, Growth Opportunity and Forecast 2019–2026. Available:

https://www.acumenresearchandconsulting.com/blockchain-in-healthcare-market, 2019 [Accessed on 31 July 2022].
[33] Deloitte, Global Life Sciences Outlook. Possibility is Now Reality, Sustaining Forward Momentum. Available: https://www2.deloitte.com/global/en/pages/life-sciences-and-healthcare/articles/global-life-sciences-sector-outlook.html, 2019 [Accessed on 31 July 2022].
[34] CVS Health, Healthcare, 2021 Full Year Guidance. Available: https://cvshealth.com/sites/default/files/media-gallery/cvs-health-earnings-press-release-2020-q4.pdf, 2021 [Accessed on 31 July 2022].
[35] Deloitte, Global Health Care Outlook. Are We Finally Seeing the Long-Promised Transformation? Available: https://www2.deloitte.com/content/dam/Deloitte/global/Documents/Life-Sciences-Health-Care/gx-health-care-outlook-Final.pdf, 2022 [Accessed on 31 July 2022].
[36] 3 Ways That Blockchain Can Help the Healthcare Industry Focus More on Patient Care and Less on Administrative Headaches, Fisher Phillips. Available: https://www.jdsupra.com/legalnews/3-ways-that-blockchain-can-help-the-1711780/, 2021 [Accessed on 31 July 2022].
[37] Protenus, Breach Barometer. Available: https://www.protenus.com/resources/2021-breach-barometer, 2021 [Accessed on 31 July 2022].
[38] D. Romano and G. Schmid, "Beyond Bitcoin: recent trends and perspectives in distributed ledger technology," *Cryptography*, vol. 5, no. 36, p. 36, 2021.
[39] Change Healthcare using Hyperledger Fabric to Improve Claims Lifecycle Throughput and Transparency. Available: https://www.hyperledger.org/learn/publications/changehealthcare-case-study, 2022 [Accessed on 31 July 2022].
[40] MELLODY – Machine Learning Ledger Orchestration for Drug Discovery. Available: https://www.melloddy.eu/, 2022 [Accessed on 31 July 2022].
[41] LOINC. Available: https://loinc.org/, 2022 [Accessed on 31 July 2022].
[42] HL7 Version 3 Product Suite. Available: https://www.hl7. org/implement/standards/product_brief.cfm?product_id=186, 2022[Accessed on 31 July 2022].
[43] HL7 Clinical DocumentArchitecture Release 2. Available: http://www.hl7.org/implement/standards/product_brief.cfm?product_id=7, 2022 [Accessed on 31 July 2022].
[44] Developing Smart Contracts with IBM Blockchain Platform Developer Tools. Available: https://cloud.ibm.com/docs/blockchain?topic=blockchain-develop-vscode, 2022 [Accessed on 31 July 2022].
[45] Hyperledger Fabric Gateway SDK for Java. Available: https://hyperledger.github.io/fabric-gateway-java/, 2022 [Accessed on 31 July 2022].
[46] AIFA. Available: https://www.aifa.gov.it/documents/20142/241052/OsMed_2019_Eng.pdf, 2022 [Accessed on 31 July 2022].
[47] T. Nakaike, Q. Zhang, Y. Ueda, T. Inagaki, and M. Ohara, "Hyperledger Fabric performance characterization and optimization using GoLevelDB benchmark," in *2020 IEEE International Conference on Blockchain and Cryptocurrency (ICBC)*, 2020.
[48] Hyperledger Caliper. Available: https://www.hyperledger.org/use/caliper, 2022 [Accessed on 31 July 2022].

Chapter 8

Challenges and future work directions in e-healthcare using blockchain technology

Dianne Scherly Varela de Medeiros[1], *Diogo Menezes Ferrazani Mattos*[1], *Natalia Castro Fernandes*[1], *Antonio Augusto de Aragão Rocha*[1], *Célio Vinicius Neves de Albuquerque*[1] *and Débora Christina Muchaluat-Saade*[1]

Abstract

Electronic healthcare (e-healthcare) emerges as a trend for enhancing traditional healthcare systems that are costly and ineffective. Besides, blockchain technology provides a distributed and secure environment for sensitive applications. This chapter discusses challenges and future work directions for digital healthcare using blockchain technology. First, the chapter introduces how blockchain technology applies to healthcare applications. Then, it discusses the main requirements for developing blockchain-based healthcare applications. In addition, the chapter presents the key challenges for applying blockchain in this sensitive-data context, emphasizing critical technological and social aspects. Next, the chapter highlights the advantages and disadvantages of applying blockchain in healthcare. Finally, it concludes with future work directions and opportunities for research on the topic.

Keywords: Blockchain; Healthcare; Patient-centric model; Healthcare supply chain

8.1 Introduction

The speed of technology growth and the globalization foster the demand for high quality health systems, facilities, and treatments. However, considering all the aspects involved in the healthcare system, such as the sensitivity inherent to the process and data, the commercial values, and its operational nature, undoubtedly, such kind of modernization requires large support of technological solutions.

[1]MidiaCom Laboratory, Universidade Federal Fluminense (UFF), Brazil

Blockchain has the potential to play a critical role in the digital transformation of the healthcare sector, with blockchain-based solutions being used in the healthcare system to exchange patient data through hospitals, medical clinics and institutions, diagnostic laboratories, health insurance companies, among others, preserving patient privacy, increasing performance, security, and transparency of sharing medical data. Blockchain-based applications can even help to identify mistakes accurately and correct procedures [1].

Blockchain technology is a current and viable solution for the healthcare industry, but this technology also creates numerous challenges that need to be solved or at least considered when coming up with blockchain-based solutions [2]. While blockchain has potential to improve efficiency and to create business opportunities, it also requires regulation, information security, and transparency. For instance, while sharing patients' electronic health records (EHRs) helps to speed and improve diagnosis, the privacy and security preservation of such data is imperative. Node scalability, data storage, and social and cultural issues such as doubts about how the technology works or resistance to structural changes in the healthcare sector are few of the challenges faced by blockchain in healthcare. Hence, it is clear that there are scenarios in which it is hard to adopt the technology either due to technical, social, or regulatory issues.

Two blockchain-based applications in healthcare that stand out are distributing EHRs and tracking supply chain information. However, challenges intrinsic to blockchain technology strongly impact applications. These challenges are scalability and the need for standardization and specific legislation. Thus, this chapter argues that blockchain technology can be disruptive to the healthcare sector by allowing the paradigm shift from healthcare professional-centric data management to a new patient-centric model. The data management paradigm shift reflects challenges for technology scalability and data privacy. Moreover, standardization and legalization of blockchain-based applications still need political and social incentives. Thus, the contribution of the chapter is five-fold:

1. surveying the most recent and relevant studies that propose advancements for blockchain-based digital healthcare;
2. highlighting the main advantages and disadvantages of using blockchain to solve the major challenges faced by the healthcare industry;
3. emphasizing the major requirements for developing blockchain-based solutions;
4. identifying key challenges for blockchain deployment in different healthcare domains;
5. envisioning future work directions, stressing the main trends for developing blockchain-based solutions in the healthcare industry.

To cover all those important aspects, the rest of this chapter is organized as follows. Section 8.2 discusses the main requirements for developing blockchain-based healthcare applications over different perspectives. Section 8.3 presents key challenges for applying blockchain to healthcare applications, emphasizing critical technological and social aspects. Section 8.4 points out benefits and drawbacks of existing blockchain-based e-health solutions, related to health data, her, and supply chain management. Section 8.5 debates on the trends and future challenges for blockchain-based healthcare applications and architectures. Finally, Section 8.6 concludes this chapter with final remarks.

8.2 Requirements for developing blockchain-based healthcare applications

Blockchain technology gained notoriety in 2009 when a person or group launched a white paper, under the pseudonym of Satoshi Nakamoto, that first presented the Bitcoin cryptocurrency [3]. From there on, uncountable blockchain solutions, including several cryptocurrencies, have been developed. The blockchain concept aims at decentralization as a security measure, and data records are distributed and shared to create a global index for all transactions in each network. It works as a public, shared, and universal ledger, which builds trust and relies on consensus in direct communication among parties without an intermediary [4].

Registering a transaction in the blockchain data structure requires that each member applies asymmetric cryptography, guaranteeing truth in stored data. This recording transaction also ensures pseudo-anonymization, given that the identities of the involved parties are hidden from the network. Transactions are grouped into blocks that must be verified through a consensus mechanism before being considered an effective transaction. There are many well-known consensus mechanisms, such as proof-of-work (PoW), proof-of-authority (PoA), proof-of-steak (PoS), to mention a few [5]. In addition to the validation, any transaction stored in the blockchain is considered safe and reliable since it also stores the cryptographic hash for assuring the chain integrity.

Blockchain technology is constantly evolving to meet application requirements. An important feature resulting from this evolution is the smart contract. This feature was first introduced in the Ethereum platform [6] and consists of a self-executing piece of code stored in the blockchain. A smart contract may be triggered by any state change or transaction registration in the blockchain and has the potential of facilitating commercial negotiation, validation, and execution without the need of a third party.

Blockchain-based solutions have widely been used to generate irrefutable computational evidences, distributedly stored, of the chronological order of transactions. The technology is well-known for its potential applications in the finance sector given its seminal solution based on Bitcoin [7]. Nevertheless, it immensely benefits a plethora of sectors due to its decentralization and intrinsic security characteristics. For instance, voting systems [8], banking and financial [9,10], monitoring of food production, and transport logistics [11,12]. In summary, blockchain can be applied where applications (i) require the contribution of multiple stakeholders and trust between parties that is complex to provide with current technology, (ii) need of reliable tracking of activity and overtime data reliability, and (iii) benefit from the removal of intermediary parties, increasing the overall system efficiency [13]. Among all industry sectors that can benefit from blockchain-based solutions, this chapter focuses on the healthcare industry, highlighting the benefits, drawbacks and main challenges for using blockchain to develop e-health applications. Figure 8.1 examines and identifies strengths, weaknesses, opportunities, and threats faced by blockchain technology in healthcare depicted as a SWOT analysis [14–16].

The healthcare industry emerges as an exciting candidate sector in which blockchain technology has the potential to play a key role, mainly due to the following key elements [14,17]:

- **Decentralization**: there is no need for an intermediary and the database system is available to anyone connected to the network that has the required access level. Monitoring, storing, accessing, and updating data can be performed on the multiple systems that are part of the network.
- **Transparency**: data recorded and stored into a blockchain is transparent to the users, meaning that all users can view the transactions performed via blockchain.
- **Immutability**: stored data cannot be modified easily, allowing stakeholders to prove with mathematical certainty that the historical data stream is exact and unmodified [13].

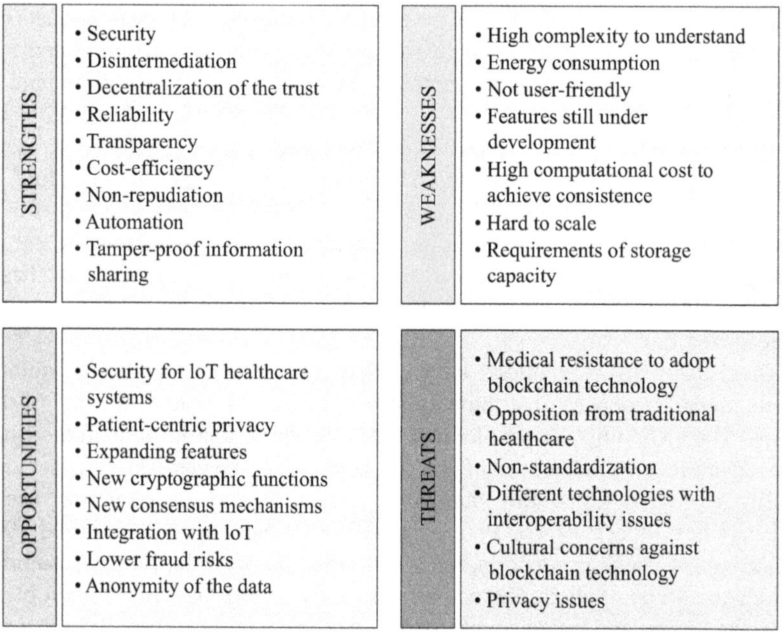

STRENGTHS
- Security
- Disintermediation
- Decentralization of the trust
- Reliability
- Transparency
- Cost-efficiency
- Non-repudiation
- Automation
- Tamper-proof information sharing

WEAKNESSES
- High complexity to understand
- Energy consumption
- Not user-friendly
- Features still under development
- High computational cost to achieve consistence
- Hard to scale
- Requirements of storage capacity

OPPORTUNITIES
- Security for IoT healthcare systems
- Patient-centric privacy
- Expanding features
- New cryptographic functions
- New consensus mechanisms
- Integration with IoT
- Lower fraud risks
- Anonymity of the data

THREATS
- Medical resistance to adopt blockchain technology
- Opposition from traditional healthcare
- Non-standardization
- Different technologies with interoperability issues
- Cultural concerns against blockchain technology
- Privacy issues

Figure 8.1 SWOT analysis of blockchain technology applied to healthcare data management

Table 8.1 Main requirements for healthcare blockchain-based applications

Requirement	Brief description
Multi-party problem	Application applied to a natural multi-party problem, interacting with multiple health professionals and institutions
Distributed trust	There is no prior trust among parties
Data consistency assurance	Blockchain ensures database integrity over time
Immutable storage	The database in the blockchain cannot be changed or deleted
High availability	Distributed databases must always be available, which is a natural characteristic of blockchains
Limited scalability and real-time processing	Reading and writing in blockchains are slower activities than in traditional databases, reducing scalability and limiting real-time processing. Also, processing delay usually varies over time
Data transparency and validation	Blockchains assure data transparency to all parties with simple validation
User privacy restrictions	Private user data cannot be stored in the blockchain as plaintext or ciphered. Applications must employ sensitive data protection tools
Identity and key management	The use of blockchains for healthcare requires a proper identity and key management among stakeholders
Variable usage costs	The use of public blockchains implies costs for every transaction in the blockchain. The developed application will require user investments
Legalization of blockchain technology	Blockchain usage for health data must be under country laws

- **Autonomy**: nodes in the network are independent and autonomous, being able to access, transfer, store, and update data safely without external intervention.
- **Anonymity**: participants' identities are anonymous, contributing to the system security and reliability.

The usage of blockchain for healthcare applications is a decision that depends on software architecture and mainly on the application's functional requirements. The following subsections discuss the main requirements to develop a healthcare application based on blockchain, which are summarized in Table 8.1.

8.2.1 Multi-party and trust

A healthcare application able to assist with patient data whenever required is a natural multi-party problem. Multiple health professionals and institutions will create a data history that helps discover patient conditions and treat future diseases. Privacy concerns also apply in this scenario, as professionals cannot share data without the previous authorization of the patient. Hence, a significant problem is creating a secure health data sharing environment. This multi-party environment requires a pre-existing trust among stakeholders to create processes for storing and fully sharing transparent data. A more realistic scenario would be to develop a trustless and transparent healthcare system that fits a blockchain structure [18].

Intermediary parties are usually involved whenever performing transactions, and operations involve multiple parties, as in supply chains. As a central intermediate agent, a trusted authority for mediating transactions and data access can provide secure processes. Still, it also becomes a single failure point that may compromise the whole system. Blockchain creates a shared infrastructure without trusted third parties and reduces costs in information sharing. Hence, this technology suits scenarios with multiple parties in which agents must trust the data recording and access, making sure that all operations follow a specific protocol as soon as a smart contract can describe this protocol. As a consequence, blockchain provides applications with the ability to remove the need to trust a specific third party, replacing this need for a distributed trust [19].

The core of the distributed trust is that the multi-party agents can develop smart contracts that comprise code to regulate interactions among mutually untrusted parties. No single party controls the system database and access control policies in a blockchain application. On the contrary, each user/agent can manage their data access, while the blockchain ensures reliability by its distributed nature.

Before developing a blockchain-based application, it is essential to check whether this kind of database suits the application requirements. Figure 8.2 shows a preliminary check to verify whether the development of a blockchain solution is justified, considering the application context [19,20]. As blockchain introduces

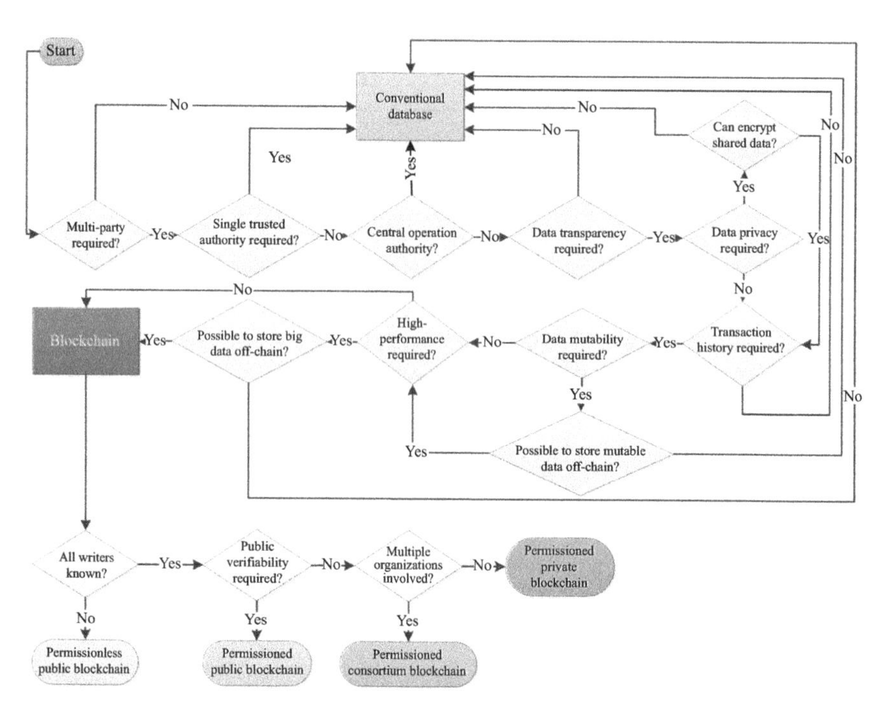

Figure 8.2 Flowchart to decide when to use the blockchain technology when developing an application

complexity to the application, a system designer should avoid it unless strictly necessary. Therefore, an application should recur to a blockchain in case it meets the following requirements [21]:

- **Database transparency**. The database needs transparently shared ledgers.
- **Distributed stakeholders**. There are distributed multiple domain orchestrators (readers and writers) with dynamic scaling and diverse stakeholders.
- **Distributed trust**. Stakeholders of different organizations that do not share trust by definition.
- **Autonomous governance**. There is a need for a low-hierarchical and autonomous governance model.
- **End-to-end service**. Stakeholders need an end-to-end service, upholding shared service level agreements (SLAs) through smart contracts.
- **Common interest among peers**: Readers and writers share a common interest in the consortium, demanding certification and auditing check-ups.

8.2.2 Storage

A conventional database and a blockchain differ in the immutability of data and the data storage structure. A traditional relational database presents data as a set of tables consisting of two-dimensional indexed arrays. Other variants include b-tree and user-defined vectors. The database control usually follows the central or distributed controller model that can securely access a hierarchical structure. Any error that corrupts a table will probably disrupt the entire service [22]. Data lake models are another variant for holding a large amount of data, which may or may not be structured. Application users can add, remove, and edit data in traditional databases and data lake cases. Transaction history is not the system's focus, and the database does not systematically provide backup, instead it should be a concern for the data administrator.

Blockchain provides storage based on the Distributed Ledger Technology (DLT) [23], which provides data copies across several nodes by design. The data model is a chain of blocks created and organized according to cryptography algorithms, ensuring that no block can ever be modified. All nodes share the list of blocks, providing much higher integrity than traditional database systems. On the other hand, such a structure does not provide fast database searches and, due to all nodes sharing all data, it is not possible to store large amounts of data. As a consequence of these characteristics, a blockchain-based application cannot rely on a massive amount of data nor state updates by changing data. Also, users must encrypt private data before adding it to the chain, as all users will have free access to read this data. Any healthcare blockchain-based application that requires a large amount of data or privacy will need to rely on traditional databases and leave only essential transactions to the blockchain.

8.2.3 Availability, scalability, and real-time processing

Traditional network application databases rely on a client–server infrastructure, where clients send queries to the database server. This client–server model does not apply in a blockchain-based application, as the blockchain is geographically distributed. In this distributed model, the presentation authority is represented by any node in the blockchain, increasing system availability [22].

Blockchain presents a peer-to-peer (P2P) network structure with several decentralized peers backing up the whole ledger. All new transactions apply data integrity protocols through consensus algorithm and cryptography. If any blockchain node is corrupted or inaccessible, data availability and integrity are not affected due to the decentralized architecture.

Although the decentralized nature of blockchain increases the number of peers providing the database service, it is crucial noticing that writing and searching for some information in a blockchain is usually slower than in a traditional database. Real-time processing in a blockchain scenario will depend on the consensus algorithm, the number of nodes in the chain, the transaction rate, the complexity of the chain's data structure, and it may depend on the chain size.

Multi-stakeholders-based healthcare systems are usually complex in sharing and processing medical data. As a consequence, blockchain-based applications should not insert unnecessary complexity and delays [18]. In this sense, many medical transactions could limit the system's scalability, mainly when proof-of-work consensus algorithms are used. For instance, mainstream public blockchains handle on average 3–20 transactions per second, while mainstream payment services handle an average of 1,700 transactions per second [19,24]. Hence, one essential requirement when developing a healthcare blockchain application is to reduce the cost and delay of the transactions by choosing a system with a less complex consensus algorithm. Consortium and private blockchain solutions usually present better performance than public blockchain deployments. Also, the application should reduce the number of required transactions as much as possible to control application delay and scalability.

In this sense, transaction throughput may limit scalability when developing a healthcare application. Some applications, like Remote Patient Monitoring (RPM), depending on its design, may lead to a high transaction rate, making it unfeasible to use blockchain technology due to both processing and amount of data generated. The choice of the consensus algorithm can involve not only a high processing time but also significant communication traffic, which can involve a subgroup of chain nodes up to the whole set of nodes [25]. Consequently, observing blockchain transaction throughput is essential when planning a healthcare application.

Lastly, another complexity and scalability issue of healthcare applications relates to smart contracts. A smart contract can facilitate programming and automation of the rule sets in a blockchain application. Hence, the choice of using blockchain when designing the application must consider the availability of smart contracts and the security and simplicity provided by the programming language and the tools available for the developer [25].

8.2.4 Data integrity, consistency, and immutability

Blockchains are considered robust due to their distributed nature and the consensus algorithm. Although robustness is a desirable characteristic, it does not provide all data required guarantees by itself, namely integrity and consistency. Consistency is a primary requirement for healthcare applications. Inconsistent or fragmented data may cause delays and higher costs to the healthcare processes [18]. In addition, it can even mislead the diagnosis. As blockchain ensures registered data integrity and immutability, it also ensures that no data will be unattempted modified by unauthorized agents.

It is essential to highlight that using blockchain only for data integrity is an expensive choice, as there are other solutions for verifiable data persistence. The use of digital signatures, for instance, is a well-known mechanism to ensure authorship and data consistency. The existence of one or more trusted parties for handling the database can assure that no data was inappropriately excluded. Blockchains become a requirement for data integrity and consistency when there are multiple parties and no global trust in access control and availability.

This data immutability and non-repudiation provided by blockchain may be a requirement for some applications and a reject criterion for others. The cryptographic hashes inserted in each block guarantee the immutability of data and transaction history. No data can be removed or edited, and the order of registered actions cannot be changed. In addition, as the whole chain is replicated across different and independent locations, any attempt to change data in one of the replicas will raise an attack interpretation by other participants, which will reject the changes.

As a consequence of this characteristic, no data can be erased, not even by data authors, which can cause practical system issues. For instance, some patient exam data need to be kept only for a short period, as it does not present critical information and is not valuable for future diagnostics. Nevertheless, using a blockchain, it is not possible to delete old records, not even for the sake of reducing database size over time. Also, other non-critical information such as personal addresses and other personal attributes cannot be edited or erased. Another example is the EHR. In this application, information exchanged among professionals and delivered to the patient cannot be altered and would benefit from blockchain characteristics. Nevertheless, there are also health professional private data containing personal notes concerning the patient in an EHR, which should be editable by that professional in a typical healthcare application.

Other real-world problems caused by immutability include exposure of private keys, which would expose patient data. In a blockchain, besides all parties having access to the whole database and seeing the private data due to the key exposure, the exposed data cannot be replaced by a new version re-cryptographed with a new key. Hence, a key leakage would permanently expose patient privacy, which is unacceptable for healthcare applications. Other practical problems are data-entry error and court orders to delete illegal content from the blockchain [19].

Data immutability is a powerful feature that turns the blockchain inflexible. If the application requires the usage of a blockchain, then its design should consider that it is not possible to roll back and, probably, additional traditional databases may be required to store high volume, private, or editable data.

8.2.5 Interoperability, data transparency, and validation

Data transparency and validation are natural characteristics of a blockchain. Any blockchain member can validate any information by simply inspecting the chain, which is fully accessible, guaranteeing transparency, and eliminating any tampering possibility [26]. On the other hand, interoperability is an issue that arises from the need to interconnect different parties in a healthcare application. Current healthcare systems lack worldwide interoperability, even though there are well known international standards [18]. Before starting a blockchain-based healthcare application, all parties must agree on data syntax and semantics to create a consistent database. Also, there must be a study on which data should be inside and outside the blockchain. In addition, application design must provide mechanisms to correlate inside and outside information to guarantee non-repudiation and validation of data outside the chain.

8.2.6 User privacy

Data privacy is one of the main criticism points of blockchain technology. The privacy setting is limited, and all participants can access all information [19]. Therefore, storing, processing, and sharing sensitive healthcare data in a public blockchain is not advisable, not even by using cryptography. Key leakages and other shortcomings such as a single-agent possessing more the 50% processing power cause privacy breaches in blockchains. As it will be discussed in Section 8.3.2.1, if an agent has more than 50% processing power in a blockchain, that agent can control, update, alter, or manipulate the data stored in the blockchain [26].

Healthcare blockchain-based applications must provide concrete secure solutions for data privacy. Both access control and storage must be carefully designed to avoid data leakage and to allow removing data whenever this situation is required, for instance, for a legal court order. The European General Data Protection Regulation (GDPR) and other similar laws around the world define strict rules concerning patient data processing and storing to protect the patients' privacy [25]. Consequently, the application will probably not provide all data immutability and full availability. The blockchain framework for healthcare applications must embody a rich set of privacy features. A few recent studies present solutions for ensuring sensitive data protection while using blockchain technology [27–30].

8.2.7 Identity and key management

A well-designed identity management (IdM) system is one of the core requirements for a healthcare application. All actors and possible actions must be identifiable, especially for legal reasons [25]. Hence, healthcare applications must provide robust authentication features and granular-level access control mechanisms. This requirement also means that all users, including patients and health professionals, will have to own and frequently use a private key.

A crucial concern is distributing, storing, and managing missing keys. First, blockchain solutions are not usually based on passwords but on private keys that a user cannot memorize. Proper devices to store and access the key become a vital requirement for the healthcare application design. Moreover, using two-factor authentication becomes desirable to avoid misleading data due to a stolen key device, for instance.

Another concern is reestablishing a key and recovering data in case a user loses his/her key. Alternatively, in another situation, in case a user is not conscious or unable to answer due to medical conditions, health professionals will need to access the private healthcare data even without the patient's explicit consent. This problem requires a solution that mixes the blockchain platform and the application layer [22] or introduces advanced shared key mechanisms that allow the blockchain consortium to restore a key when the user loses or cannot present his key [30].

8.2.8 Technology acceptance and costs

One primary concern about blockchain in healthcare is user acceptance, as this technology is complex in both understanding and usage. Studies on user acceptance models for different technologies show that users usually adopt less complex and easy-to-use technologies [26]. In this sense, there is a cultural resistance to move habitual healthcare processes, accessed through paperwork or intuitive digital systems, to the "unknown" and complex blockchain technology. The cultural shift is then a significant challenge [18].

Another concern about technology acceptance is the open nature of the public blockchains. Sensitive data is stored and available for all blockchain nodes. Even using cryptography, this issue raises concerns about data privacy and should be further analyzed before the extensive implementation of blockchain-based healthcare systems [26].

Finally, operational costs are another concern about technology acceptance. For healthcare applications, systems should present low and predictable operational costs, which restrains the blockchain platforms to ones with predictable and stable transaction fees/cost [25].

8.2.9 Legalization of blockchain technology

Legalization is a vital requirement when dealing with private health data. In this sense, blockchain technology still lacks rules and regulations worldwide. For instance, blockchain-based applications in the government domain remain limited. Considering that governments in most countries offer and control health services, it raises legal implications for health applications [26].

Laws, such as the European General Data Protection Regulation (GDPR), bring concerns on how personal data is defined and protected, implying restrictions to business models that rely on blockchain. One particular concern is the inability to remove data from the blockchain, as those laws usually allow users to delete their data whenever they please, as specified in the "Right to be Forgotten" and the "Right of Erasure" [20]. It is essential to notice that health records are generally defined as exceptional cases in these regulations, as the patient information concerns the patient and healthcare professionals and facilities. Hence, the restrains of the use of blockchain in healthcare should be evaluated in each country, considering local laws.

In a broader view, there are still challenges in defining policies for the collaboration of stakeholders in a complex ecosystem that already presents regulatory frameworks, such as in the context of healthcare applications [18]. For instance, initiatives such as the Health Insurance Portability and Accountability Act (HIPAA) [31] are developing standards to preserve the privacy of health records.

8.3 Key challenges for healthcare applications using blockchain technology

Traditional healthcare systems are costly regarding financial, time, and computing resources. Intermediary parties are frequently necessary to perform transactions [18], such as the authorization and payment of a medical exam. An intermediary party is needed due to the lack of trust between the parties involved in the healthcare system [18], but the intermediary's presence incurs time and financial cost. The lack of trust also jeopardizes data storing and sharing, hindering the availability of complete health data. Systems that store and collect such data rely on the interaction between patient and physician due to a doctor-centric health delivery paradigm. In this paradigm, the physician assesses the patient's health status based on information derived from medical examination and previously stored information present in the patient's Health Record (HR) [32]. Healthcare is tailored around the physician and the systems fail to take advantage of the stored data. Patients' health data are scattered across the various healthcare systems, which makes providing data security and reliability a complex task. Such scattering also results in lack of critical data availability, which often incurs on healthcare failing to provide the best treatment to the patients. Moreover, there is a long and tedious bureaucracy involved with accessing healthcare [33].

Blockchain helps improving and securing the manipulation of patients' EHRs and supporting system interoperability among healthcare stakeholders [18]. Blockchain can provide utmost privacy while ensuring that only authorized users can manipulate data stored in a permanent record of information regarding patients' health [13]. Blockchain's decentralization characteristic protects data from tampering by individual stakeholders and external parties [13]. The implementation of blockchain-based healthcare can also minimize the cost and resources due to the current intermediary parties present in the healthcare delivery architecture, as it helps preserving trust among stakeholders [18]. Moreover, blockchain potentially allows access to a massive volume of anonymized health data, enabling new research on healthcare while simultaneously protecting patients' privacy, serving as a protocol to connect stakeholders to data without requiring a trusted third party [13]. Hence, blockchain has the potential to provide less complex processes for the healthcare delivery system and reduce the delays introduced by the presence of trusted intermediary parties [18].

Nevertheless, implementation of blockchain-based healthcare requires healthcare delivery to move from a doctor-centric to a patient-centric paradigm. The patient-centric paradigm evolves the relationship between patients and medical providers. Their tasks become to go beyond clinical treatment of patients and to get involved in their emotional, physical, mental, social, and financial situations, toward empathizing on a deeper level to provide better medical treatment.

Moreover, this paradigm drives the patients to actively participate in their own medical care delivery [32]. The healthcare delivery becomes ubiquitous to prevent the deterioration of patients' health condition. Hence, in the patient-centric paradigm, patients are under the ubiquitous monitoring of a healthcare service that provides context to personalize healthcare delivery to each individual [32]. To achieve the patient-centric healthcare delivery paradigm, a set of new principles, such as collaboration, openness, integrity, and health data co-creation by patients are required [13]. Blockchain has the potential to help achieve these principles and provide personalized, reliable, and secure healthcare [14].

In this context, there are several challenges faced by healthcare. Blockchain can provide a unified and secured view, and exchange of EHRs through a flexible, adaptable, agile, and secure infrastructure with high performance and low latency that can help to solve such challenges. Nevertheless, applications based on the blockchain technology can be costly and complex to develop. Blockchain is an emerging technology surrounded by uncertainty about real deployments. As such, there is a lack of evaluation of healthcare blockchain-based applications in real scenarios. Moreover, patients' lack of awareness about the technology and regulatory challenges poses barriers for the wide adoption of blockchain-based healthcare. The following sections discuss the challenges related to healthcare that can be addressed by blockchain technology, as well as the challenges related to the use of this technology in the healthcare industry.

8.3.1 Healthcare challenges

The HRs are traditionally stored on paper and scattered across the healthcare system. Management of patients' health data in the healthcare industry is still very inefficient, being a complex task to gather the data to extract information that can be used to improve patients' treatment and reduce medical errors [34]. Hence, primal challenges of the healthcare industry are to safely retrieve health data generated by the daily business activities and properly manage this data. Health information is a valuable asset, and it is obtained from sensitive personal data. Due to its value, health data databases are a prime target for ransomware and other cyberattacks. Hence, the security of such data is paramount. Moreover, healthcare data must be protected from unauthorized access and manipulation. Nevertheless, healthcare systems face several limitations, such as patient's privacy and data integrity, quality and accuracy.

Patients' health data scattered across several stakeholders' healthcare systems is often inaccessible to other stakeholders and, usually, the systems used lack interoperability due to standardization issues. It is imperative to provide a nationwide standardization for the entire healthcare ecosystem. The lack of coordinated data management and sharing due to the nonexistence of interoperability can cause fragmentation of healthcare information, leading to incomplete patients' health data [18]. Even if healthcare providers have access to all scattered information in the traditional healthcare system, these data still can be enriched to provide more efficient treatment. In the patient-centric paradigm, healthcare providers should have access to patients medical and socioeconomic data at any given time and anywhere, allowing suiting the treatment in a personalized and ubiquitous model. Fraud is another common problem in the healthcare industry, either due to billing or supply chain issues. Missing payment and medical drug disappearance happen frequently.

In this context, the main healthcare challenges can be divided into the following domains: (i) health data management, (ii) supply chain management, (iii) standardization, billing and payment, and (iv) context-awareness and personalization. Table 8.2 summarizes the discussion about healthcare challenges, highlighting the challenges in each domain and how blockchain can help to tackle them. The following sections discuss each domain challenges in detail.

8.3.1.1 Health data management

Health data management is one of the greatest challenges in the healthcare industry, since it involves data security, privacy, storage and sharing. When immersed in a patient-centric paradigm, the volume of data generated is massive, as it is obtained from a multitude of sources, such as wearables and medical appointments. The massive volume of heterogeneous data is stored in centralized silos, making data management and sharing a challenging task. Even though important information can be extracted from this data, the distributed storing in non-interoperable data silos hinders data access. Moreover, the data are usually stored in a non-standardized format. These characteristics make the health data challenging to understand, share, and process.

Sharing data is an essential task in healthcare to allow transferring the clinical data of clinical practitioners' patients to the concerned authority for a quick follow-up. The data transfer between authorized entities is always a concern and data storing must meet strict security requirements and strong measures for transparency and accountability due to patients' privacy [14]. Nevertheless, over the last decade, more than 2,100 data breaches have happened only in the healthcare sector of the United States of America, affecting millions of records. It is expected that between 60% and 80% of data breaches are unreported. Unauthorized access or disclosure accounts for 34% of the healthcare data breaches, while 47% come from hackers or various technology incidents, such as outdated systems and ineffective security measures. The most popular targets are healthcare and finance sectors, with healthcare accounting for 15% of the attacks*. Losing access to critical information, such as health records, costs millions of dollars to victim organizations. Therefore, it is paramount to increase the security of healthcare systems. Legislation to protect personal information should be taken into account [13].

The proper management of health data and the sharing and retrieval of such data in a safe, secure, and scalable way allows gathering recommendations from clinical specialists [14]. As such, the clinical communication becomes more efficient, enabling healthcare systems to create holistic views of patients and improve

*All information retrieved from https://techjury.net/blog/healthcare-data-breaches-statistics/. Accessed on January 31, 2022.

Table 8.2 Summary of healthcare challenges

Domain	Challenge	How blockchain helps
Health data management	Data silos are centralized and non-interoperable	Provides a secure common distributed database-like structure
	Information is scattered across stakeholders' systems, causing databases to be incomplete and out-of-date	Allows shared write access for several parties while guaranteeing data integrity due to consensus mechanisms
	Data format is not standardized	Allows the deployment of smart contracts that define the data format and serve as a common interface to store and retrieve data, enabling uniform portability
	Several security breaches exist (e.g., unauthorized access)	Provides means for authentication, confidentiality, and accountability
Supply chain management	Several security breaches exist (e.g., counterfeiting, pilfering and fraud)	Improves the traceability of products, and consequent identification of the distribution step in which the drug was tampered, due to the immutability of records stored in the blockchain
Standardization billing and payment	Lack of interoperability among several stakeholders is common	Allows the deployment of smart contracts that define the data format and serve as a common interface to store and retrieve data, enabling uniform portability
	Billing inaccuracies and billing related frauds are common	Enables multifaceted protection system and prevents forged transactions due to the decentralized nature and consensus mechanisms
	Claiming bills takes a long time	Allows the development of new payment models with transparent information about transactions, taking inputs from different sources without tampering with any information
Context-awareness and personalization	Management of patients' profiles, medical and socioeconomic data is hard	Provides data integrity and means for authentication, confidentiality and accountability
	Information is scattered across stakeholders' systems, causing databases to be incomplete and out-of-date	Allows shared write access for several parties while guaranteeing data integrity due to consensus mechanisms
	Patients' data need to be available at all times	Provides a secure common distributed database-like structure

the quality of care and treatments. Hence, it is imperative to provide efficient health data management for diagnosis and in combined clinical decision making [14]. Failing to provide this service efficiently results in security breaches and shattered information. Frequently, patients do not have full access to their medical information, scattered over multiple institutions. Shifting this paradigm to a patient-centered information sharing should allow patients to access and control their data. Moreover, better treatment outcomes are expected to be achieved by ensuring complete and up-to-date health information availability for the authorized involved participants [13,14]. EHRs, however, are not designed to handle lifetime records among multiple institutions. As patients' data are scattered among these institutions, patients do not have easy access to their past medical data. Patients' data can be compromised during storage and sharing among several entities, posing a threat to the patients' health status. Maintaining patients' historical medical data up to date can be lifesaving and, thus, it is an imperative feature to ensure accurate diagnostics and effective treatment [14].

Blockchain technology is suitable for any kind of application that uses digital data and requires authentication and consensus about data integrity, as well as shared write access for several parties. Blockchain enables health providers to create an integrated system for storing and accessing EHRs, giving patients the control over their medical data and, thus, contributing to establish the patient-centric healthcare delivery paradigm. Moreover, blockchain keeps medical history from not being available or well-maintained due to discontinuity and unavailability of previous health records. It adds more transparency to the process of data sharing in the form of data transactions, as it maintains a distributed ledger of hard-to-temper data among all involved entities within the network, making all parties aware of transactions [18]. Due to blockchain characteristics, many researchers claim this technology to be a good choice for maintaining EHRs, as it can provide control over data, data provenance auditing, and secured data trailing [14]. Hence, blockchain keeps important medical data safe and secure [35], providing a trustable and secure solution for recordkeeping the complete information about the health status of each patient. As such, patients do not need to repeat the same information several times to different people involved in their healthcare [18]. A recent study proposes reliable storage solution that preserves privacy for users and enables perpetual recordkeeping of important data [36]. Other authors study how to protect patient data from unauthorized access using blockchain to manage authentication, confidentiality, accountability, and data sharing for EHRs [37,38].

8.3.1.2 Supply chain management

In the healthcare industry, supply chain management is complex, and any failure can affect the well-being of a patient. The supply chain is vulnerable and permeated by security breaches that allow fraudulent attacks to happen, as the supply chain involves a number of moving parts and people [14]. Medical drugs counterfeiting and pilfering are common in the healthcare industry, causing huge financial loss to pharmaceutical companies. Raw materials, compounds, or components can also be deviated from their original delivery path.

Blockchain provides a safe and secure platform to eliminate the several security issues existing in the supply chain process, helping to prevent frauds due to the introduction of data transparency and improvement on the traceability of products. The immutability of records stored in the blockchain hinders the manipulation of the stored data to commit fraud. Hence, the blockchain technology improves the ability to identify the origin and authenticity of medical products, consequently enhancing supply chain security. As the blockchain can also help to track the production and location of drugs at any given time, enhancing the integrity of the overall supply chain process, the technology can also help on the traceability of counterfeiting drugs [14]. It is possible to track the distribution of drugs from its production to its delivery to the patient, and identify in which step of the supply chain the drug was tampered. As such, it is possible to ensure that the medical drug provided to the consumers is legitimate and the patient can verify the authenticity of such drug before buying it [39].

8.3.1.3 Standardization, billing, and payment

The healthcare industry faces the challenge of recording and storing information in such a way that sharing it securely across several different applications is possible. The format is usually not portable across systems due to lack of uniformity and compatibility between these systems. There is no nationwide interoperability between current healthcare systems. Inefficient interoperability hinders the flow of information in the healthcare system, which can jeopardize the management of public health threats and the provisioning of adequate health treatments. Patients need to have easy access to their health records and physicians must be able to access all the scattered information of their patients easily to make more accurate diagnostics and provide efficient treatment. Hence, maintaining the interoperability among various stakeholders is a core issue and it is imperative to create universal standards to achieve such interoperability [18]. Billing inaccuracies and frauds are also major issues in current healthcare systems. Traditional patient billing systems are complex and exposed to billing-related frauds. Moreover, when an insurance company is involved in the payment process, several weeks can pass before bills can be claimed.

Blockchain empowers EHR interoperability for the current healthcare systems [18], but as a new technology it will face standardization challenges as discussed later in this section. This technology enables uniform portability and multifaceted protection systems such that blockchain-based systems can help in the reconciliation of records and activities as well as in curbing fraud. Decentralized record-keeping helps payment processing and prevents forged transactions. The maturity of blockchain in the financial sector can be used to create a new payment model based on blockchain for the healthcare industry. Such a model creates a sense of trust and brings transparent information about transactions. As such, mistrust between organizations and payers could be avoided [40]. A blockchain-based payment model can also help insurance companies to claim bills faster and decrease the time and cost involved in the billing process [18]. Faster claims can happen due to the possibility of taking inputs from a variety of different sources without tampering with any information [40]. Hence, blockchain-based payment solutions facilitate the billing process, providing reliable medical insurance storage, and ensuring the primitiveness and verifiability of stored data while simultaneously providing high credibility to users [40].

8.3.1.4 Context-awareness and personalization

In the patient-centric healthcare delivery paradigm, the patient must be the core entity in the healthcare ecosystem. The concept of precision medicine, as discussed in [41], involves prevention and treatment strategies that take individual variability into account. In order to achieve that, personalized healthcare is needed, which can only be provided if there is a chain of collaboration between the various healthcare entities. Patients form a social network composed by a group of individuals that serve predefined roles to deliver the healthcare service, for instance, medical professionals and caregivers. The patients' social network must have access to medical as well as socioeconomic information about the patient, as both medical and socioeconomic factors influence an individual's health status. The data can be obtained from biosensors, context-aware sensors, global positioning system (GPS), mobile terminals, among others [32]. Socioeconomic data cover environmental, social, and communal factors, and play an essential role in the development of health management programs for a population [42].

The richness of information that can be obtained from medical and socioeconomic factors can help healthcare professionals to determine how patients are managing chronic health problems and diseases, and to create patients' profiles. With such profiles, it becomes easier to provide more personalized healthcare treatments. Access to patients' profiles, medical and socioeconomic data must be managed and the system providing such information must guarantee that the profile data is synchronized when it is required. As such, a profile management system is required for the efficient support and provisioning of personalized healthcare service. Current healthcare systems are not able to provide such service and deliver healthcare mainly in healthcare facilities. The patient-centric paradigm demands this service to be provided anywhere and anytime. This can only be achieved by applying ubiquitous computing to bridge the gap between the virtual and physical worlds, integrating seamlessly information and communication technologies with people in their daily lives. Context-awareness is indispensable for personalized healthcare, as it guides the customization of such service. Hence, being able to apply ubiquitous computing upon patients' context data acquired anytime and anywhere is a crucial step for the establishment of patient-centric healthcare delivery [32].

Blockchain is well-known for its capacity to provide data integrity due to the immutability characteristic of the technology. Moreover, it is widely applied in several domains to provide access control to sensitive data, guaranteeing privacy, and secure access. Hence, blockchain technology can solve the challenges of data sharing, privacy, and immutability to provide personalized ubiquitous healthcare delivery to patients, speeding up the availability of actionable socioeconomic and behavioral health data into EHRs.

8.3.2 Blockchain adoption challenges

Blockchain has the potential to create a new foundation for economic and social systems, but several obstacles exist, hindering quick and wide adoption of the technology on the various industry sectors. Focusing on the healthcare industry, patients lack knowledge about the technology and are often unwilling to share and store their personal medical information due to fear of having data leaked [18]. At the same time, application developers that build services based on blockchain technology need to tackle scalability, as the number of participants in the healthcare industry is huge. The effort of building a new healthcare system, more secure and digitized, is not worth if standardization and regulatory laws are not taken into account. Figure 8.3 summarizes the main challenges that are discussed in this section related to the adoption of blockchain technology in the healthcare industry, considering technical, regulatory, and social issues.

8.3.2.1 Technical challenges

Blockchain-based applications are being developed and refined over time. Even though this technology brings many benefits and opportunities to several industry sectors, it must also address its own set of challenges [14]. Implementing applications based on blockchain technology eliminates the need for a third party to carry out a transaction. As blockchain allows all nodes to verify records in the blockchain architecture rather than a single trusted third party, the data is prone to potential privacy and security risks. Data privacy will not be intact as all nodes access the data transmitted by one node. Patients need to choose a proxy that can access their information or medical history on their behalf if an emergency happens. The proxy can allow people to access the same patient's records, creating a massive threat to data privacy and security. Blockchain networks, mainly proof-of-work-based ones, are also prone to breach cryptographic security through the attack known as the

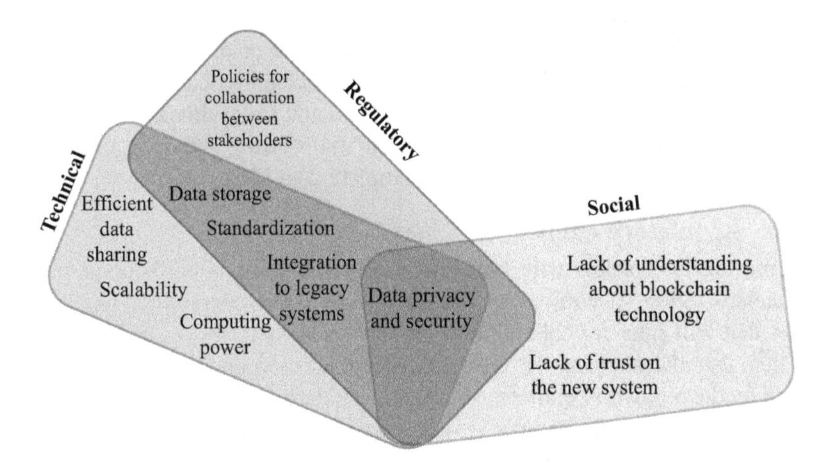

Figure 8.3 Key challenges for adoption of blockchain in the healthcare industry according to technical, regulatory, and social issues

51% attack. In this attack, a group of miners owns more than 50% of the computing power of the blockchain network and, therefore, can decide which blocks are added to the blockchain [14].

Blockchain adoption and integration will incur great costs to institutions and needs to tackle the lack of scalability, the non-standardization, the potential for information decay, the storage of graphical data, the integration to legacy systems, the interconnection of nodes to supply the necessary computing power to create blocks, and the time needed to verify and add new transactions to the blockchain [43], to name a few challenges. Blockchain design concerns recording and processing transaction data with a limited scope and, thus, it does not intend to store a high volume of data. Nevertheless, the huge number of participants in a blockchain-based healthcare system produces a vast volume of medical transactions that can limit the system's scalability. As such, the more blockchain technology expands into the healthcare sector, the more evident the storage challenges become. The healthcare industry holds a large volume of data to be processed daily, which hardens effective data sharing. There is also a trade-off between the volume of transactions and the available computing resources, which hinders scalability. Patients' data are not shared by multiple parties and blockchain does not solve the interoperability problem if each stakeholder deploys its own non-standardized solutions.

Blockchain is still seen as a complimentary technology to the legacy systems, instead of a substitute. This is because most of healthcare data would be kept off the blockchain, and each block would have information to retrieve such health data. Data sharing, although essential for blockchain-based healthcare systems, can raise security and privacy concerns. Only authorized entities should access and store the data [36]. The blockchain-based healthcare solutions do not need a trusted third party to carry out transactions. All nodes can potentially access the records in a blockchain architecture, making data prone to potential privacy and security risks [14]. An authorization mechanism is required to select the entities that are allowed to access the information. The authorized entities are usually assumed to be trustworthy, but this is not true in a real settlement and can result in huge data privacy and security threat. Nevertheless, this is not a problem tied only to blockchain-based solutions. Tackling security and privacy is complex and involving data and the healthcare system in high security mechanisms can result in hurdles for transferring the data, causing data consumers to have access to limited or incomplete data [14].

8.3.2.2 Regulatory challenges

Replacing legacy systems with blockchain-based systems is not economical [44] and standardization is required to facilitate overall interoperability. Defining the policies that will consider collaboration of various stakeholders to draft a complete ecosystem that also accounts for the existing regulatory framework is a challenge to be faced by the regulatory entities. Moreover, standards must be defined to preserve privacy of patients' health records [45], so that clinical data can be safely, securely, and successfully exchanged between healthcare system participants. Such an exchange requires trustworthiness and a healthy collaboration among the involved entities. As potential constraints faced in this process, the nature of clinical data, sensitivity, data sharing agreements, procedures, complex patient matching algorithms, ethical policies, and governing rules can be highlighted [14]. When each entity creates its own blockchain solution, making the several blockchains from various communicating providers and services talk between them seamlessly and appropriately is a challenge that creates hindrances in the effective sharing of data [14].

The storage of healthcare data, although a technical challenge, also results in a regulatory challenge, as there is a need to find a common ground and agree to store all information in a nationally recognized standard format. As such, organizations need to align on a framework for defining which data, size, and format can be submitted to the blockchain, and which must be stored off the blockchain [46]. Well-authenticated and certified standards are required for accomplishing an international standardization. These predefined standards are helpful to assess the size, data nature, and format of the information exchanged among blockchain applications. The standards need to scrutinize the shared data as well as serve as precautionary safety measures [14]. Nevertheless, regulatory actions are slow, leading to the development of incompatible back-end systems, and fragmented shared medical information.

8.3.2.3 Social challenges

All participants of the healthcare system are used to follow legacy healthcare processes. Healthcare system participants usually do not share information among them, meaning that data are kept in uncommunicating data silos. Accepting and adopting a new technology that is completely different from traditional methods is not easy. Hence, patients and healthcare providers are afraid of sharing sensitive data and using a new system developed over a technology that they do not clearly understand. The healthcare sector is slowly moving towards digitization, but it still requires wide clinical validation of the blockchain-based approaches before adopting them as the new state-of-the-art. The paperwork bureaucracy involved in the healthcare sector could be reduced by adopting blockchain solutions, but such adoption will still take time and effort, as the blockchain-based solutions currently offered are relatively untrusted [14]. Thus, the major social challenge is changing the mindset and behavior of people towards data sharing and using a new technology.

8.4 Benefits and drawbacks of existing blockchain-based healthcare applications

The following sections discuss existing blockchain solutions proposed in the literature related to health data and EHR considering patient-centric approaches, scalability analysis and supply chain management for the usage of blockchain in healthcare systems.

8.4.1 Health data and EHR

One of the primary concerns when dealing with EHRs is that those data belong to patients, but they are entirely controlled by healthcare institutions [47]. Another concern relates to identity management (IM), as IM enhances trust and the privacy over EHR systems [48]. IM for EHR storing and consulting systems is usually centralized, which introduces a single point of failure and an access bottleneck for the whole system [49]. Hence, although there are different blockchain-based proposals for storing and sharing EHRs [37,49–51], there are some flaws still unsolved.

Dubovitskaya *et al.* store EHRs in a permissioned blockchain, which is interesting for keeping user data privacy, but they fail on providing fine-grained access control to patient information [49].

Azaria *et al.* propose MedRec, a processing-intensive consensus mechanisms that is based on a public blockchain [37]. MedRec [37] shares EHRs among patients, hospitals, and physicians using open standards of health data. Also, MedRec use Ethereum smart contracts to control access to healthcare history, being able to recover fragmented medical records of different healthcare institutions at the cost of high computational processing.

Magyar proposes an approach based on storing health records in a blockchain [52], while Zang *et al.* use a pervasive social network that allows users to share medical sensors data that is registered in the blockchain [53].

Guo *et al.*'s work associates blockchains and an attribute-based signature scheme based on multiple authorities for creating a distributed EHR system [54]. The proposed scheme introduces a patient capability to securely manage Personal Health Records (PHR). Nevertheless, this facility also brings a performance cost, as it creates an overhead for signing the transaction by multiple authorities. This proposal also suffers from confidentiality issues concerning data stored in the blockchain.

In a similar way, Dang *et al.* also use attribute-based signatures to guarantee privacy and confidentiality of EHR in fog and cloud environments [55]. They analyze the use of fog computing for storing and securing EHRs [55].

Yue *et al.* focus on providing a coarse-grained privacy control [51]. The proposed system uses mobile phones to interact with an access control gateway that controls access to blocks in the blockchain. Nevertheless, that gateway does not control transactions.

There are many different EHR sharing systems that use blockchain technology to improve the quality of medical information retrieval [28,37,56–58]. Medicalchain's platform [57] is a Health Data Marketplace accessed by MedTokens. Five hundred million MedTokens were issued and sold in 2018. In their solution, the patient controls access of records to doctors, for instance during telemedicine consultation. MedTokens can also be used to pay for doctor consultations. Patients may also grant to researchers access of their personal health records, in return for MedTokens.

Similar to Medicalchain, the MedChain solution [56] uses two distinct token types: (1) external tokens, named MedCoins, to provide access control and privacy; and internal tokens, named Record Tokens, to provide a map of the distributed patient record, by adding hashes to the blockchain. MedChain records can include health data in various formats such as plain text, digital images or database objects.

MediBChain's major concern is to provide privacy [28]. It uses Pseudonymity through the use of cryptographic public keys to protect the patient's identity. It implements a blockchain-based patient centric healthcare data management system.

AuditChain [58] provides multilevel access control for patients, physicians, nurses and hospital administrators for the management of EHRs. It implements smart contracts in Hyperledger Fabric[†] [59]. Transaction's digital signature uses public key cryptography and serves as a virtual token for access control. Query scripts are required to find the patient's health records.

Uddin *et al.* [60] present a role-based access control that is deployed by a patient-centric agent with the drawback of requiring high processing power for data encryption and authentication of the patient's smartphone. Its protocol is prone to the man-in-the-middle attack. A security and privacy decentralized sharing of health records proposal [61] uses attribute-based encryption to ensure privacy.

Zhang *et al.* [50] discuss the granularity of access control of data stored in a blockchain. The authors claim that block-level permission does not offer appropriate permission control on e-health data. They propose layered access to the encrypted content within the blockchain. Thus, their proposal introduces a delay on retrieving data, since the access control layer validates every access.

Oliveira *et al.* [38] develop an EHR approach whose access control is patient-centric. Their approach relies on both Public Key Infrastructure (PKI) and blockchain technology. The key idea is to inherit the trust on the authenticity provided by the PKI and the integrity and the accountability provided by the blockchain. Unlike previous work, they propose a secure distributed EHR that requires a computationally simple infrastructure, and provides fine-grained access control, with little overhead. Table 8.3 summarizes the benefits and drawbacks of existing blockchain-based EHR proposals.

8.4.2 Scalability analysis

Parket *et al.* [62] attempted to experimentally demonstrate the limits of using blockchain for sharing medical data. To do that, they created a private network based on Ethereum technology with a hospital and 300 patients. At the end of the experiment, they concluded that it was fundamental: to minimize the amount of data actually recorded on the blockchain, improve data privacy and consider transaction costs. The work was based on the Ethereum network, in which each transaction has a custom in GAS[‡]. That cost is proportional to the size of the stored data and inversely proportional to the write priority. If the official Ethereum network was used, the cost of a transaction would currently be at least around 10 US cent. Even though it is not a high cost, a question still remains to determine whether the cost must be absorbed by clients or by healthcare entities. Attention should also

[†]Available at https://www.hyperledger.org/use/fabric. Accessed on February 15, 2022.

[‡]According to Ethereum documentation, GAS is the fuel that allows it to operate. GAS fees are paid in Ethereum's native currency, ether (ETH). GAS prices are denoted in gwei—each gwei is equal to 0.000000001 ETH. Available at https://ethereum.org/en/developers/docs/gas/. Accessed on February 14, 2022.

Table 8.3 Benefits and drawbacks of existing blockchain-based EHR proposals

Proposal	Key feature	Benefits	Drawbacks
Dubovitskaya *et al.* [49]	Permissioned EHR blockchain	User data privacy	Limited access control to patient data
Azaria *et al.* (MedRec) [37]	Public EHR blockchain	Access control by Ethereum smart contracts	High computational processing
Guo *et al.* [54]	Distributed EHR system	Attribute-based signature for secure PHR management	Signing costs and confidentiality issues
Dang *et al.* [55]	Fog and cloud EHR system	Attribute-based signature for privacy and confidentiality	Signing costs and limited environment
Yue *et al.* [51]	Gateway-based EHR system	Coarse-grained privacy control	Limited transaction control
Medicalchain [57]	Health Data Marketplace	Patient-centric privacy control	Complex MedToken Management
Medchain [56]	Health Data Marketplace	Patient-centric privacy control and Map of distributed records	Complex MedCoin and Record Token Management
MediBChain [28]	Patient centric EHR	Public key pseudonymity	Complex key management
AuditChain [58]	Multilevel access control	Virtual token: transaction's digital signature	Complex query scripts
Uddin *et al.* [60]	Patient centric agent	Role-based access control	High processing cost
Zhang *et al.* [50]	Layered access control	Fine-grained permission control	Long validation and retrieval delays
Oliveira *et al.* [38]	Patient-centric Distributed EHR	Fine-grained access control	PKI-based

be drawn to the unpredictable fluctuation in the value of the digital currency of Ethereum, the ether (ETH), which may eventually make the gas price higher, increasing the cost to execute transactions. The work of Parker *et al.* also drew attention to the fact that health data is sensitive data that cannot be left open on blockchain networks, it must be protected by cryptographic mechanisms.

The work of McGhin *et al.* [63] summarizes the technological challenges for the real implementation of blockchain-based health information systems. The text brings two interesting points: (i) it presents a compilation of software projects that support healthcare applications using blockchain and (ii) it lists vulnerabilities aggregated to healthcare systems through the use of the blockchain. The latter, in other words, reminds us that health information systems are already subject to severe security requirements and, when using blockchain, new vulnerabilities must be taken into account, such as, for example, the susceptibility to attacks [64].

Altogether, McGhin *et al.* [63] highlight nine initiatives, including MedRec [37], Gem Health Network [65], OmniPHR [44], PSN [53], Virtual Resources [66], Context-driven Data Logging [67], MedShare [68], Trial and Precision Medicine [69], and Healthcare Data Gateways [51]. The authors draw attention to the fact that all studies have scalability and/or security limitations. For example, the MedRec project [37], which is one of the most cited articles on blockchain applications for healthcare, deals with the use of blockchain for the auditable storage of the patient's medical interactions history. The architecture, based on Proof-of-Work, allows mediation of access permission to data, but, according to McGhin *et al.* [63], the MedRec prototype does not satisfactorily guarantee the anonymity of individual data and makes it possible, through forensic techniques, to discover the identity of patients and service providers.

8.4.3 Supply chain management

Secure product origin tracking complemented by blockchain technology has found its implementation possibilities in many sectors, especially pharmaceutical supply chains [15]. Blockchain technology provides reliable information about the origins, treatment, storage, and validity of pharmaceutical products from their production to the final consumer. A blockchain-based pharmaceutical supply chain system simultaneously helps patients and healthcare staff read information from many products and record new information about them. Blockchain allows faster and more accurate identification of the source of problems, followed by recalls and other measures.

Clauson *et al.* [70] discuss challenges and opportunities of the use of blockchain technology to enhance supply chain management in healthcare. They focus on the pharmaceutical supply, medical device and supplies, Internet of Things (IoT) for Healthcare, and public health sectors. As a drawback, they indicate that most blockchain initiatives remain in the pilot phase, which highlights the need for further evaluation and alignment with policy mechanisms.

Musamih *et al.* [71] present yet another prototype approach for drug traceability in the healthcare supply chain. The paper describes an Ethereum blockchain-based approach in which smart contracts are used to provide provenance. The history of all transactions is stored in the blockchain in a secure, immutable fashion. Smart contract security analysis was performed highlighting integrity, accountability, authorization, availability, non-repudiation and resiliency to the man-in-the-middle attack. Furthermore, SmartCheck and the Oyente tool were used to detect smart contracts' code vulnerabilities. Among the limitations, there are issues related to data privacy, scalability, interoperability, and efficiency. As a final comment, the proposed smart contracts are tailored for the healthcare supply chain, nevertheless they can be clearly adapted to other types of supply chains.

Similarly, Omar *et al.* [72] propose a healthcare supply chain (HCSC) process based on the Ethereum network able to connect manufacturers, Group Purchasing Organizations (GPO), distributors, and providers. The proposal uses the blockchain to develop decentralized storage to promote transparency, streamline communication with stakeholders, and reduce procurement timelines. The authors developed a generic framework based on smart contracts with detailed algorithms depicting various interactions among HCSC stakeholders to automate the GPO contract

Table 8.4 Benefits and drawbacks of existing healthcare supply chain (HSC) management proposals

Proposal	Key feature	Benefits	Drawbacks
Clauson *et al.* [70]	Survey on challenges and opportunities for blockchain-based HSC management	Analysis of pharmaceutical, medical, IoT and public health sector supplies	Most initiatives remain in the pilot phase
Musamih *et al.* [71]	Drug traceability proposal	Ethereum blockchain-based approach with smart contracts	Data privacy, scalability, interoperability and efficiency issues
Omar *et al.* [72]	HSC to promote transparency among stakeholders	Ethereum blockchain-based approach with smart contracts	Transaction fees
Sadri *et al.* [73]	Blood donation supply chain	Ethereum blockchain-based approach with smart contracts (role-based)	All blood supply stages must join the blockchain

process. Tests with the author's implementation showed that the proposal is economically feasible because it depends on a minimal transaction fee.

Another challenging issue for healthcare sectors is an effective blood supply chain system. In healthcare, blood is essential in medical treatment; however, there is a lack of efficient supply chains to cope with this essential resource. Such supply chain system is important once blood relevant information needs to be traceable at each stage of the blood supply, from the donors to the blood recipient. All of this with the necessary transparency, auditability, privacy, security, and trustworthy. Sadri *et al.* [73] propose a blockchain-based solution using Ethereum platform that leverages traceability in the blood donation supply chain (BDSC). In their solution, smart contracts are used to ensure the access roles, which guarantees traceability and security of information in the BDSC ecosystem. Table 8.4 summarizes the benefits and drawbacks of existing supply chain management proposals.

8.5 Future work directions

Blockchain applications and usage for healthcare tend to become extensive in a near future [74]. The leading research and application interests lay into two main categories: applications of interest and blockchain architecture trends. The following sections discuss the most promising applications of blockchain in healthcare and architectural trends of blockchain solutions.

8.5.1 Healthcare blockchain-based application trends

The most promising blockchain-based applications in healthcare aim to provide supply chain transparency, patient-centric EHRs, smart contract settlements, medical staff credential verification, security for remote health monitoring, or data provenance. This section discusses those healthcare applications, and lists current deployments as follows.

8.5.1.1 Supply chain transparency

Supply chain management requires attention to long-lasting products as well as items that require special equipment and facilities for sale, storage, and distribution [75]. A significant open issue across the healthcare industry is ensuring the provenance of medical products to confirm their authenticity and ensure product tracking. Tracking items with a blockchain-based system at every stage of the supply chain, from manufacturing to consumption, provides transparency to customers who can assert the provenance of health-related products. It is one of the main applications of blockchain technology in industry, especially in drug-developing markets, preventing drug counterfeiting and incorrect drug consumption. Accurate tracking is essential to avoid counterfeit and fraudulent medicines and drugs. Transparency across the entire supply chain is vital for medical devices, such as remote health monitoring systems.

iSolve[§] is an application that controls the lifecycle of health items based on a blockchain architecture. iSolve is based on the Advanced Digital Ledger Technology (ADLT), an enterprise blockchain solution. ADLT aims to complement existing systems and processes, ensure data provenance, bypass data silos, create a practical platform, and create an interoperable and high-performance environment. ADLT uses blockchain to track, verify and preserve the entire logistical movement for drug lifecycle and drug supply chain development in the biopharmaceutical and healthcare industry. ADLT can control the lifecycle from production to distribution via blockchain and, thus, control the drug provenance, reducing the likelihood of fraudulent re-labeling or date change.

8.5.1.2 Patient-centric electronic health records

Healthcare systems in all countries and regions are struggling with the problem of data silos, which means that patients and their healthcare providers have an incomplete view of medical histories. EHRs are critical in improving intelligence, quality, user experience, and costs related to the healthcare system. The EHR stores the patient's private information about diagnosis and treatments. The private information is susceptible but often shared among untrusted peers such as healthcare providers, pharmacies, patient families, and other physicians. The management of health records challenges privacy while delivering data to authorized peers. The patient must control the information sharing with only the peers he/she trusts. Moreover, the lack of interoperability between the health provider and hospital systems makes sharing information among peers in the healthcare system problematic. Thus, records are fragmented in local databases, which prevents the patient from having a consolidated health history. Likewise, patients fail to control where their health data are and which peers access them [38]. A potential solution for sharing health data is a blockchain-based system in which references to patient

[§]Available at https://isolve.io/. Accessed on January 13, 2022.

record entries are stored in the blockchain, such as through hash functions, and the content of the entries is stored externally.

The Medicalchain[||] company uses blockchain technology to make a patient-centric EHR and maintain a single accurate report of user data. The application's primary goal is to put the patients in control of their medical data, giving them the power to share the single most comprehensive version of their health history with all the organizations in their medical network. Medicalchain uses blockchain technology to securely manage health records for a collaborative and intelligent approach to healthcare. Medicalchain deals with hospital discharge summaries, which summarize the treatment and essential follow-up care. Hospitals need to make sure these documents are free of errors and quickly process them to release patients to the following line. Sending documents across administrative boundaries may require a written request, and problems with repetitive information, fraud, and inaccessible information are prevalent. Medicalchain offers a digitized solution that guides physicians through a structured discharge procedure that decreases errors and deletions and speeds up reviews.

8.5.1.3 Smart contract settlements in healthcare

Smart contracts execute computer code directly on a blockchain [76]. Smart Contracts present a straightforward "if-then" logic and can automatically trigger another contract or execute a transaction when meeting predefined criteria. As an example, the autonomous execution of smart contracts makes them ideal for obtaining informed patient consent to participate in clinical trials, and the immutability of shared data supports the validity of consent. Since smart contracts execute automatically and patient's consent is available to all chain members immediately after signing, smart contracts are a crucial tool for performing clinical trials over data in a blockchain data structure.

Chronicled** relies on smart contracts to provide a more reliable, efficient, and automatic supply chain. It provides tools and protocols for multi-party supply chain ecosystems using decentralized blockchain networks. The decentralized network has developed trust boundaries and enforces business rules across the organization while keeping data private. Besides, Chronicled launched the Mediledger Project[††]. a leading consortium of pharmaceutical companies that run a blockchain-based compliance protocol as an accounting system attributed to pharmaceutical supply chain security, privacy, and efficiency.

8.5.1.4 Medical staff credential verification

Similar to tracking the provenance of health items, blockchain technology may also track the experience of medical professionals, where trusted medical institutions and healthcare organizations record the credentials of their employees, helping to facilitate the hiring process for healthcare organizations. In this way, ProCredEx[‡‡]

[||]Available at https://medicalchain.com/. Accessed on January 13, 2022.
**Available at https://www.chronicled.com/. Accessed on January 14, 2022.
[††]Available at https://www.mediledger.com/. Accessed on January 14, 2022.
[‡‡]Available at https://hashedhealth.com/blockchain-healthcare-solutions/professional-credentials-exchange/. Accessed on January 14,2022.

is an accreditation exchange service sponsored by the Hashed Health organization[§§], which stores and verifies accreditation information, such as including licensing information, for medical doctors [77]. The exchange may integrate several healthcare stakeholders, such as hospitals or clinics, who need credential information to work. ProCredEx provides fast access to selected data, released and available for members' exchanging.

8.5.1.5 Security and interoperability for remote health monitoring

A significant trend in digital health is the adoption of remote health-monitoring solutions, in which sensors measure patients' vital signs and help healthcare professionals to have more visibility about patients' health, enabling more proactive and preventive care. Although the trend is current, security is still a big open issue in the healthcare Internet of Things (IoT) because ensuring that patient data are private, secure, and not tampered with to create false information is still an open question. In cases where an IoT device is the trustworthy item to initiate an emergency, for example, alerting an elderly caregiver that the person under care has suffered a fall or a heart attack, it is crucial to assure that the IoT device is resistant against attacks that create fake information or disrupt the service [77].

Polkadot[||||||] is a decentralized blockchain platform that creates parallel chains, called parachains. Parachains are custom blockchains that become part of the global Polkadot main network. A Parachain is a data structure similar to a blockchain and is verifiable by validators on the Polkadot Relay Chain. Parachains inherit security mechanisms from the relay chain and communicate with other Parachains through the Cross-chain Message Passing (XCMP) protocol. Parachains are maintained by a cluster node that acts as a supervisor, retaining all transactional information, producing new blocks, and passing them on to the Relay Chain for inclusion in the main network.

Cosmos[***] is an ecosystem of interconnected blockchain-based applications and services. Cosmos relies on the Tendermint protocol, in which independent blockchains called zones plug into the larger blockchain, the Cosmos network. All zones are connected to the Cosmos Hub and communicate secure message passing protocols. Zones maintain their consensus mechanism due to a sharding consensus scheme.

8.5.1.6 Health data provenance

Data provenance refers to a historical record of the data source and lifecycle, showing how data items are stored, accessed, and processed by whom and for what purpose [78]. Data provenance increases transparency and enhances data integrity. Blockchain technology offers immutable record storage, and smart contracts may act as a data controller to verify data origin before storing it.

[§§]Available at https://hashedhealth.com/. Accessed on January 14, 2022.
[||||||]Available at https://polkadot.network/. Accessed on January 14, 2022.
[***]Available at https://cosmos.network/. Accessed on January 14, 2022.

MediBloc[†††] is a decentralized health information ecosystem built on top of blockchain technology for patients, healthcare professionals, and researchers [79]. MediBloc allows patients to track and record all their health-related details, such as doctor visits and health records, on its blockchain platform. Separate from aggregating data, it also assigns data ownership to patients. When sharing information, patients are rewarded with native currency MED tokens, which can be used to pay for MediBloc partner products and services, such as pharmaceutical purchases and insurance.

8.5.2 Blockchain architectural trends

Blockchain is a recent and constantly evolving technology. However, there are still open issues inherent to the technology critical to the development of healthcare applications. The primary healthcare-related issue to the development of blockchain technology is related to legal issues; reliability on off-chain stored items; the chain immutability; the technology scalability; the fitness of consensus mechanisms to the healthcare ecosystem requirements; and issues related to the social acceptance and standardization of the technology [78].

8.5.2.1 Legal issues

Healthcare is a sector with unique requirements associated with security and privacy due to legal requirements to protect patients' medical information [63]. Legal issues are risks that harden the design and deployment of healthcare blockchain applications. Legal systems are geographically bounded, whereas blockchain technology intrinsically distributes applications and, thus, crosses those boundaries. Participants in blockchains cross different law boundaries. Hence systems may be subject to a foreign jurisdiction [80]. Legal issues related to smart contracts are also crucial for adopting healthcare applications based on smart contracts. As previously mentioned, the European General Data Protection Regulation (GDPR), for instance, states that citizens have the "right to be forgotten," which conflicts with the built-in immutability of the blockchain data structure [78].

8.5.2.2 Reliability of off-chain resources

In blockchain technology, off-chain resources refer to data or computation structurally external to the blockchain network [81]. Off-chain blockchain systems enable information processing and management through distributed software architecture, in which the blockchain network interacts with off-chain resources. The main motivations for using off-chain blockchain systems are improving scalability, reducing data storage requirements, and improving data privacy, which are critical issues to enable wider blockchain adoption. Off-chain processing is an essential component of data governance that explores the interaction between on-chain and off-chain storage and computing. Besides, smart contracts usually require receiving information or parameters from resources off-chain. Hence, trusted third

[†††]Available at https://medibloc.com/en/. Accessed on January 14, 2022.

parties retrieve information off-chain and then send the information to the blockchain system at given times. Nevertheless, the introduction of a third party introduces a point of failure.

Blockchain applications with extensive medical data, images, documents, and laboratory results need significant storage space. As each node in the blockchain network, the system keeps a copy of all records, which may cause a lack of storage space for current blockchain technology. Furthermore, current blockchain platforms are transaction-based, and therefore the databases used for the technology tend to overgrow. The growth of databases hampers records' search and access speed, which is highly inappropriate for medical applications where fast access to data is crucial. Thus, off-chain resources are mandatory for storing voluminous data, while the main blockchain only stores references and summaries of the original data [78].

8.5.2.3 Chain immutability

Immutability is a fundamental and overriding feature of blockchains. Once committing a transaction, its outcome cannot be changed no matter who or why. Nevertheless, when considering smart contracts and automatic code execution in blockchain, there is the implication that immutability incurs that, in case of errors made in the code, there is no possibility of correcting the code or previous erroneously generated outcomes. Consequently, if there are changes in circumstances, such as changing laws and regulations, there is no simple way to change a smart contract, requiring extensive and potentially expensive reviews of the smart contract code by experts before deploying it on a blockchain. Remarkably, there is a need for extensive testing of the smart contract code prior to its commitment to the blockchain. It is also important to point out that the irreversibility of the blockchain impeaches reversing smart contracts or transactions already committed to the chain. Thus, if a node of the blockchain network is attacked or misused to report erroneous data, it is impossible to reverse the on-chain effects.

8.5.2.4 Blockchain technology scalability

Scalability is a primary concern for many blockchain networks. Scalability is related to block size, high transaction volume, and the number of nodes [82,83]. Block size refers to the maximum block capacity to hold transactions. If the capacity exceeds, the network rejects the block. High transaction volume refers to transactions taking place in real-time. When a blockchain system is used as a database to store patient profile data, records replicate across all blockchain participant nodes in a large-scale healthcare scenario. Consequently, an overflow of information occurs, leading to a high data storage volume in healthcare institutions. Existing solutions require the blockchain to generate a high amount of transactions to be processed and connected to the network, resulting in performance degradation. Another architectural issue is that data synchronization capability may be lower than the transaction rate required for enrollment in the blockchain system. Besides, the difficulty of running blockchain-based applications would increase significantly with the growth in the number of members or patients, as it requires an increase in the computational power of the entire blockchain infrastructure. Solving the scalability issue demands extensive research focusing on increasing the number of transactions per second by blockchain platforms in the future. Transaction verification depends on the consensus mechanism used and, thus, scalability also depends on consensus mechanisms.

8.5.2.5 Fitness of consensus mechanisms

The consensus is to maintain blockchain networks' security, scalability, and decentralization mechanisms. The global consistency of the blockchain is due to the correctness and the consistency of its distributed data structure. As a distributed asynchronous system, the blockchain relies on a consensus mechanism to keep its shared data consistent [5,84]. Several consensus mechanisms, either probabilistic or deterministic consensus algorithms, suit different blockchain deployment scenarios. Probabilistic consensus mechanisms for blockchain, such as those based on Proof of Work (PoW) and Proof of Authority (PoA), sacrifice the deterministic agreement convergence in favor of partition tolerance and availability to provide network scalability. Deterministic consensus mechanisms either adopt a loose-consistency model, which does not resist misbehavior attacks, or adopt a byzantine fault-tolerant model, which incurs a high rate of message exchanging. Thus, it is still an open issue to search for the most fitted consensus mechanism for each healthcare application and provide interoperability between blockchain platforms that run different consensus mechanisms.

8.5.2.6 Social acceptance and standardization

Healthcare-related blockchain applications are subjected to all standardization and regulation of the healthcare sector. It is mandatory to have an adequate and standardized way of storing data from the patients' health records and exchanging information between the different systems. Although blockchain technology may play the role of a standardized interface among different systems, the healthcare stakeholders must convey in a single and interoperable blockchain ecosystem. International standards authorities must create well-authenticated and certified standards. Standards help assess the size, data origin, and information format exchanged in blockchain applications. Standards are not intended to examine shared information for privacy but should also provide barriers to avoid information leakage.

Blockchain technology is still progressing, but meets societal disbelief, and requires a cultural shift. Accepting the technology that differs from traditional working methods is never easy. Although a solution for recording transactions through blockchain and smart contracts has as one of its goals to provide greater transparency to the operations of the healthcare system, blockchain adoption will only be successful if it counts on the trust of patients, physicians, and other healthcare stakeholders. It is an open issue, mainly because most participants in the system are technology-naïve users. Most patients are uncomfortable with sharing their data with multiple participants, and the odds of cultural resistance ensue [78].

8.6 Conclusion

Blockchain technology is still evolving and faces social challenges such as cultural shifts. Accepting and adopting a technology completely different from traditional working methods is never easy. Precision healthcare that demands healthcare personalization is a very important issue in the healthcare sector where patient-centric EHRs are crucial. To achieve this new healthcare paradigm, a set of new principles as collaboration by multiple parties, data integrity, and health data co-creation by patients should be considered. This chapter discussed how and why blockchain has potential to help achieve these principles and provide personalized, reliable, and secure healthcare. An important blockchain technological challenge refers to scalability. Blockchain-based applications cannot currently rely on a huge amount of data. Any healthcare blockchain-based application that requires a large amount of data or privacy will need to combine the use of traditional databases and leave only essential transactions to the blockchain. As discussed in this chapter, a potential solution for sharing health data is a blockchain-based system in which references to patient record entries are stored in the blockchain, and the content of the entries is stored externally. Due to blockchain technology low adoption rate in the healthcare industry, the technology and policies offered are relatively unreliable. As a crucial social challenge, patients may be unwilling to share and store their personal health information in blockchain-based systems due to fear of having data leaked. As blockchain is a recent technology, government regulations and more successful blockchain-based use-case deployments to prove its usefulness in a broader scale are expected for healthcare information systems. Main trends in the development of blockchain applications in healthcare show that the near future applications are developing integrated, distributed, private EHR with solid encryption and patient-centric secure data management architectures. Blockchain technology in healthcare still requires several improvements in technologies and procedures to meet privacy and security standards. Prospects are promising regarding the development of new blockchain-based applications and the development of new blockchain-inspired system architectures, such as distributed ledgers and blockchain-based healthcare payment channel networks.

References

[1] A. Haleem, M. Javaid, R. P. Singh, R. Suman, and S. Rab, "Blockchain technology applications in healthcare: an overview," *International Journal of Intelligent Networks*, vol. 2, pp. 130–139, 2021.

[2] D. A. Pustokhin, I. V. Pustokhina, and K. Shankar, *Challenges and Future Work Directions in Healthcare Data Management Using Blockchain Technology*, pp. 253–267. Singapore: Springer Singapore, 2021.

[3] S. Nakamoto, *Bitcoin: A Peer-to-Peer Electronic Cash System, Technical Report*, 2008.

[4] E. Jesus, V. Chicarino, C. Albuquerque, and A. Rocha, “A survey of how to use blockchain to secure internet of things and the stalker attack,” *Security and Communication Networks*, vol. 2018, pp. 1–27, 2018.

[5] G. R. Carrara, L. M. Burle, D. S. V. Medeiros, C. V. N. Albuquerque, and D. M. F. Mattos, “Consistency, availability, and partition tolerance in blockchain: a survey on the consensus mechanism over peer-to-peer networking,” *Annals of Telecommunications*, vol. 75, pp. 163–174, 2020.

[6] V. Buterin, “A next-generation smart contract and decentralized application platform,” *white paper*, vol. 3, no. 37, 2014.

[7] U. Mukhopadhyay, A. Skjellum, O. Hambolu, J. Oakley, L. Yu, and R. Brooks, “A brief survey of cryptocurrency systems,” in *14th Annual Conference on Privacy, Security and Trust (PST)*, pp. 745–752, 2016.

[8] K. Patidar and S. Jain, “Decentralized e-voting portal using blockchain,” in *10th International Conference on Computing, Communication and Networking Technologies (ICCCNT)*, pp. 1–4, IEEE, 2019.

[9] Q. K. Nguyen, “Blockchain – a financial technology for future sustainable development,” in *3rd International Conference on Green Technology and Sustainable Development (GTSD)*, pp. 51–54, IEEE, 2016.

[10] M. Osmani, R. El-Haddadeh, N. Hindi, M. Janssen, and V. Weerakkody, “Blockchain for next generation services in banking and finance: cost, benefit, risk and opportunity analysis,” *Journal of Enterprise Information Management*, vol. 34, pp. 884–899, 2020.

[11] F. Tian, “An agri-food supply chain traceability system for China based on rfid & blockchain technology,” in *13th International Conference on Service Systems and Service Management (ICSSSM)*, pp. 1–6, IEEE, 2016.

[12] L. Koh, A. Dolgui, and J. Sarkis, “Blockchain in transport and logistics–paradigms and transitions,” *International Journal of Production Research*, vol. 58, no. 7, pp. 2054–2062, 2020.

[13] M. A. Engelhardt, “Hitching healthcare to the chain: an introduction to blockchain technology in the healthcare sector,” *Technology Innovation Management Review*, vol. 7, pp. 22–34, 2017.

[14] A. A. Siyal, A. Z. Junejo, M. Zawish, K. Ahmed, A. Khalil, and G. Soursou, “Applications of blockchain technology in medicine and healthcare: challenges and future perspectives,” *Cryptography*, vol. 3, no. 1, p. 3, 2019.

[15] D. Dujak and D. Sajter, *Blockchain Applications in Supply Chain*, pp. 21–46. Cham: Springer International Publishing, 2019.

[16] E. J. De Aguiar, B. S. Faiçal, B. Krishnamachari, and J. Ueyama, “A survey of blockchain-based strategies for healthcare,” *ACM Comput. Surv.*, vol. 53, pp. 1–27, 2020.

[17] S. Namasudra, P. Sharma, R. G. Crespo, and V. Shanmuganathan, “Blockchain-based medical certificate generation and verification for IoT-based healthcare systems,” *IEEE Consumer Electronics Magazine*, vol. 12, no. 2, pp. 83–93, 2023.

[18] T. Kumar, V. Ramani, I. Ahmad, A. Braeken, E. Harjula, and M. Ylianttila, “Blockchain utilization in healthcare: key requirements and challenges,” in

20th International Conference on e-Health Networking, Applications and Services (Healthcom 2018), pp. 1–7, 2018.
[19] S. K. Lo, X. Xu, Y. K. Chiam, and Q. Lu, "Evaluating suitability of applying blockchain," in *2017 22nd International Conference on Engineering of Complex Computer Systems (ICECCS)*, pp. 158–161, 2017.
[20] F. Zemler and M. Westner, "Blockchain and GDPR: application scenarios and compliance requirements," in *Portland International Conference on Management of Engineering and Technology (PICMET)*, pp. 1–8, 2019.
[21] R. V. Rosa and C. E. Rothenberg, "Blockchain-based decentralized applications for multiple administrative domain networking," *IEEE Communications Standards Magazine*, vol. 2, no. 3, pp. 29–37, 2018.
[22] W. Yang, E. Aghasian, S. Garg, D. Herbert, L. Disiuta, and B. Kang, "A survey on blockchain-based internet service architecture: requirements, challenges, trends, and future," *IEEE Access*, vol. 7, pp. 75845–75872, 2019.
[23] M. J. M. Chowdhury, M. S. Ferdous, K. Biswas, *et al.*, "A comparative analysis of distributed ledger technology platforms," *IEEE Access*, vol. 7, pp. 167930–167943, 2019.
[24] V. Sivaraman, S. B. Venkatakrishnan, K. Ruan, *et al.*, "High throughput cryptocurrency routing in payment channel networks," in *17th USENIX Symposium on Networked Systems Design and Implementation (NSDI 20)*, Santa Clara, CA, pp. 777–796, USENIX Association, February 2020.
[25] C. C. Agbo and Q. H. Mahmoud, "Comparison of blockchain frameworks for healthcare applications," *Internet Technology Letters*, vol. 2, no. 5, p. e122, 2019.
[26] S. Hakak, W. Z. Khan, G. A. Gilkar, M. Imran, and N. Guizani, "Securing smart cities through blockchain technology: architecture, requirements, and challenges," *IEEE Network*, vol. 34, no. 1, pp. 8–14, 2020.
[27] M. Puppala, T. He, X. Yu, S. Chen, R. Ogunti, and S. T. C. Wong, "Data security and privacy management in healthcare applications and clinical data warehouse environment," in *IEEE-EMBS International Conference on Biomedical and Health Informatics (BHI)*, pp. 5–8, 2016.
[28] A. Al Omar, M. Rahman, A. Basu, and S. Kiyomoto, "MediBchain: a blockchain based privacy preserving platform for healthcare data," *Lecture Notes in Computer Science*, vol. 10658, pp. 534–543, 2017.
[29] G. Zyskind, O. Nathan, and A. S. Pentland, "Decentralizing privacy: using blockchain to protect personal data," in *IEEE Security and Privacy Workshops*, pp. 180–184, 2015.
[30] L. Axon, M. Goldsmith, and S. Creese, "Chapter 8 – Privacy requirements in cybersecurity applications of blockchain," in *Blockchain Technology: Platforms, Tools and Use Cases* (P. Raj and G. C. Deka, eds.,) vol. 111 of *Advances in Computers*, pp. 229–278, New York, NY: Elsevier, 2018.
[31] P. Edemekong and M. Haydel, "Health Insurance Portability and Accountability Act (HIPAA)," StatPearls. Treasure Island (FL): StatPearls Publishing, 2019.

[32] M.-A. Fengou, G. Athanasiou, G. Mantas, I. Griva, and D. Lymberopoulos, *Towards Personalized Services in the Healthcare Domain*, pp. 417–433. New York, NY: Springer New York, 2013.

[33] C. Pirtle and J. Ehrenfeld, "Blockchain for healthcare: the next generation of medical records?," *Journal of Medical Systems*, vol. 42, p. 172, 2018.

[34] R. Hillestad, J. Bigelow, A. Bower, *et al.*, "Can electronic medical record systems transform health care? Potential health benefits, savings, and costs," *Health Affairs*, vol. 24, no. 5, pp. 1103–1117, 2005.

[35] C. C. Agbo, Q. H. Mahmoud, and J. M. Eklund, "Blockchain technology in healthcare: a systematic review," *Healthcare*, vol. 7, no. 2, p. 56, 2019.

[36] H. Li, L. Zhu, M. Shen, F. Gao, X. Tao, and S. Liu, "Blockchain-based data preservation system for medical data," *Journal of Medical Systems*, vol. 42, p. 141, 2018.

[37] A. Azaria, A. Ekblaw, T. Vieira, and A. Lippman, "MedRec: using blockchain for medical data access and permission management," in *2nd International Conference on Open and Big Data (OBD 2016)*, pp. 25–30, 2016.

[38] M. T. de Oliveira, L. H. A. Reis, R. C. Carrano, *et al.*, "Towards a blockchain-based secure electronic medical record for healthcare applications," in *IEEE International Conference on Communications (ICC)*, May 2019.

[39] M. Uddin, "Blockchain medledger: hyperledger fabric enabled drug traceability system for counterfeit drugs in pharmaceutical industry," *International Journal of Pharmaceutics*, vol. 597, p. 120235, 2021.

[40] L. Zhou, L. Wang, and Y. Sun, "MIStore: a blockchain-based medical insurance storage system," *Journal of Medical Systems*, vol. 42, pp. 149–149, 2018.

[41] F. S. Collins and H. Varmus, "A new initiative on precision medicine," *The New England Journal of Medicine*, vol. 372, pp. 793–795, 2015.

[42] L. S. Hughes, R. L. Phillips, J. E. DeVoe, and A. W. Bazemore, "Community vital signs: taking the pulse of the community while caring for patients," *The Journal of the American Board of Family Medicine*, vol. 29, no. 3, pp. 419–422, 2016.

[43] W. J. Gordon and C. Catalini, "Blockchain technology for healthcare: facilitating the transition to patient-driven interoperability," *Computational and Structural Biotechnology Journal*, vol. 16, pp. 224–230, 2018.

[44] A. Roehrs, C. A. da Costa, and R. da Rosa Righi, "OmniPHR: a distributed architecture model to integrate personal health records," *Journal of Biomedical Informatics*, vol. 71, pp. 70–81, 2017.

[45] M. A. Bouras, Q. Lu, F. Zhang, Y. Wan, T. Zhang, and H. Ning, "Distributed ledger technology for ehealth identity privacy: state of the art and future perspective," *Sensors*, vol. 20, no. 2, 2020.

[46] E. Karafiloski and A. Mishev, "Blockchain solutions for big data challenges: a literature review," in *17th International Conference on Smart Technologies*, pp. 763–768, 2017.

[47] M. Lesk, "Electronic medical records: confidentiality, care, and epidemiology," *IEEE Security Privacy*, vol. 11, pp. 19–24, 2013.

[48] G. Dolera Tormo, F. Gomez Marmol, J. Girao, and G. Martinez Perez, "Identity management – in privacy we trust: bridging the trust gap in ehealth environments," *IEEE Security and Privacy*, vol. 11, pp. 34–41, 2013.

[49] A. Dubovitskaya, Z. Xu, S. Ryu, M. Schumacher, and F. Wang, "Secure and trustable electronic medical records sharing using blockchain," in *AMIA '17*, vol.2017, pp. 650–659, 2017.

[50] X. Zhang and S. Poslad, "Blockchain support for flexible queries with granular access control to electronic medical records (EMR)," in *IEEE International Conference on Communications (ICC)*, pp. 1–6, May 2018.

[51] X. Yue, H. Wang, D. Jin, M. Li, and W. Jiang, "Healthcare data gateways: found healthcare intelligence on blockchain with novel privacy risk control," *Journal of Medical Systems*, vol. 40, p. 218, 2016.

[52] G. Magyar, "Blockchain: Solving the privacy and research availability tradeoff for ehr data: a new disruptive technology in health data management," in *IEEE Neumann Colloquium (NC)*, November 2017.

[53] J. Zhang, N. Xue, and X. Huang, "A secure system for pervasive social network-based healthcare," *IEEE Access*, vol. 4, p. 9239–9250, 2016.

[54] R. Guo, H. Shi, Q. Zhao, and D. Zheng, "Secure attribute-based signature scheme with multiple authorities for blockchain in electronic health records systems," *IEEE Access*, vol. 6, pp. 11676–11686, 2018.

[55] L. Dang, M. Dong, K. Ota, J. Wu, J. Li, and G. Li, "Resource-efficient secure data sharing for information centric e-health system using fog computing," in *IEEE International Conference on Communications (ICC)*, pp. 1–6, May 2018.

[56] J. Sandgaard and S. Wishstar, "Medchain," 2018. White Paper. Available on: http://medchain.us/doc/Medchain Whitepaper v1.0.pdf.

[57] A. Albeyatti, "Medicalchain," 2018. White Paper. Available on: https://medicalchain.com/Medicalchain-Whitepaper-EN.pdf.

[58] J. Anderson, *Securing, Standardizing, and Simplifying Electronic Health Record Audit Logs through Permissioned Blockchain Technology*. Ph.D. dissertation, Dartmouth College, 2018.

[59] D. Agrawal, S. Minocha, S. Namasudra, and A. H. Gandomi, "A robust drug recall supply chain management system using hyperledger blockchain ecosystem," *Computers in Biology and Medicine*, vol. 140, p. 105100, 2022.

[60] M. A. Uddin, A. Stranieri, I. Gondal, and V. Balasubramanian, "Continuous patient monitoring with a patient centric agent: a block architecture," *IEEE Access*, vol. 6, pp. 32700–32726, 2018.

[61] L. Chen, W.-K. Lee, C.-C. Chang, K.-K. R. Choo, and N. Zhang, "Blockchain based searchable encryption for electronic health record sharing," *Future Generation Computer Systems*, vol. 95, pp. 420–429, 2019.

[62] Y. R. Park, E. Lee, W. Na, S. Park, Y. Lee, and J.-H. Lee, "Is blockchain technology suitable for managing personal health records? Mixed-methods study to test feasibility," *Journal of Internet Medical Research*, vol. 21, no. 2, p. e12533, 2019.

[63] T. McGhin, K.-K. R. Choo, C. Z. Liu, and D. He, "Blockchain in healthcare applications: research challenges and opportunities," *Journal of Network and Computer Applications*, vol. 135, pp. 62–75, 2019.

[64] D. K. Tosh, S. Shetty, X. Liang, C. A. Kamhoua, K. AKwiat, and L. Njilla, "Security implications of blockchain cloud with analysis of block withholding attack," in *17th IEEE/ACM International Symposium on Cluster, Cloud and Grid Computing*, 2017.

[65] M. Mettler, "Blockchain technology in healthcare: the revolution starts here," in *IEEE 18th International Conference on e-Health Networking, Applications and Services (Healthcom)*, pp. 1–3, 2016.

[66] M. Samaniego and R. Deters, "Hosting virtual IoT resources on edge-hosts with blockchain," in *IEEE International Conference on Computer and Information Technology (CIT)*, p. 116–119, 2016.

[67] M. Siddiqi, S. T. All, and V. Sivaraman, "Secure lightweight context-driven data logging for bodyworn sensing devices," in *5th International Symposium on Digital Forensic and Security (ISDFS)*, pp. 1–6, 2017.

[68] Q. Xia, E. B. Sifah, K. O. Asamoah, J. Gao, X. Du, and M. Guizani, "MeDShare: trust-less medical data sharing among cloud service providers via blockchain," *IEEE Access*, vol. 5, p. 14757–14767, 2017.

[69] Z. Shae and J. J. Tsa, "On the design of a blockchain platform for clinical trial and precision medicine," in *IEEE 37th International Conference on Distributed Computing Systems (ICDCS)*, p. 1972–1980, 2017.

[70] K. A. Clauson, E. A. Breeden, C. Davidson, and T. K. Mackey, "Leveraging blockchain technology to enhance supply chain management in healthcare: an exploration of challenges and opportunities in the health supply chain," *Blockchain in Healthcare Today*, vol. 1, pp. 1–12, 2018.

[71] A. Musamih, K. Salah, R. Jayaraman, *et al.*, "A blockchain-based approach for drug traceability in healthcare supply chain," *IEEE Access*, vol. 9, pp. 9728–9743, 2021.

[72] I. A. Omar, R. Jayaraman, M. S. Debe, K. Salah, I. Yaqoob, and M. Omar, "Automating procurement contracts in the healthcare supply chain using blockchain smart contracts," *IEEE Access*, vol. 9, pp. 37397–37409, 2021.

[73] S. Sadri, A. Shahzad, and K. Zhang, "Blockchain traceability in healthcare: blood donation supply chain," in *23rd International Conference on Advanced Communication Technology (ICACT)*, pp. 119–126, 2021.

[74] S. Namasudra, G. C. Deka, P. Johri, M. Hosseinpour, and A. H. Gandomi, "The revolution of blockchain: state-of-the-art and research challenges," *Archives of Computational Methods in Engineering*, vol. 28, pp. 1497–1515, 2021.

[75] S. M. Hosseini Bamakan, S. Ghasemzadeh Moghaddam, and S. Dehghan Manshadi, "Blockchain-enabled pharmaceutical cold chain: applications, key challenges, and future trends," *Journal of Cleaner Production*, vol. 302, p. 127021, 2021.

[76] D. S. V. Medeiros, N. C. Fernandes, and D. M. F. Mattos, "Smart contracts and the power grid: a survey," in *1st Blockchain, Robotics and AI for Networking Security Conference (BRAINS)*, pp. 41–47, IEEE, 2019.

[77] V. Dhillon, T. Xu, and C. Parikh, "Blockchain enabled tracking of physician burnout and stressors during the covid-19 pandemic," *Frontiers in Blockchain*, vol. 3, p. 10, 2021.

[78] S. N. Khan, F. Loukil, C. Ghedira-Guegan, E. Benkhelifa, and A. Bani-Hani, "Blockchain smart contracts: applications, challenges, and future trends," *Peer-to-Peer Networking and Applications*, vol. 14, pp. 2901–2925, 2021.

[79] H. S. A. Fang, "Commercially successful blockchain healthcare projects: a scoping review," *Blockchain in Healthcare Today*, vol. 4, pp. 1–8, 2021.

[80] C. Esposito, A. De Santis, G. Tortora, H. Chang, and K.-K. R. Choo, "Blockchain: a panacea for healthcare cloud-based data security and privacy?," *IEEE Cloud Computing*, vol. 5, no. 1, pp. 31–37, 2018.

[81] K. Miyachi and T. K. Mackey, "hOCBS: a privacy-preserving blockchain framework for healthcare data leveraging an on-chain and off-chain system design," *Information Processing & Management*, vol. 58, no. 3, p. 102535, 2021.

[82] A. A. Mazlan, S. Mohd Daud, S. Mohd Sam, H. Abas, S. Z. Abdul Rasid, and M. F. Yusof, "Scalability challenges in healthcare blockchain system – a systematic review," *IEEE Access*, vol. 8, pp. 23663–23673, 2020.

[83] M. T. Oliveira, G. R. Carrara, N. C. Fernandes, *et al.*, "Towards a performance evaluation of private blockchain frameworks using a realistic workload," in *22nd Conference on Innovation in Clouds, Internet and Networks and Workshops (ICIN)*, pp. 180–187, 2019.

[84] M. T. Oliveira, L. H. Reis, D. S. Medeiros, R. C. Carrano, S. D. Olabarriaga, and D. M. Mattos, "Blockchain reputation-based consensus: a scalable and resilient mechanism for distributed mistrusting applications," *Computer Networks*, vol. 179, Article no. 107367, 2020.

Index

CPSIA information can be obtained
at www.ICGtesting.com
Printed in the USA
JSHW011738220523
42081JS00002B/5

9 781839 536021